Customer Loyalty
in Third Party Logistics Relationships

Contributions to Management Science

K. Zwerina
Discrete Choice Experiments in Marketing
1997. ISBN 3-7908-1045-2

G. Marseguerra
Corporate Financial Decisions and Market Value
1998. ISBN 3-7908-1047-9

WHU Koblenz – Otto Beisheim Graduate School of Management (Ed.)
Structure and Dynamics of the German Mittelstand
1999. ISBN 3-7908-1165-3

A. Scholl
Balancing and Sequencing of Assembly Lines
1999. ISBN 3-7908-1180-7

E. Canestrelli (Ed.)
Current Topics in Quantitative Finance
1999. ISBN 3-7908-1231-5

W. Bühler/H. Hax/R. Schmidt (Eds.)
Empirical Research on the German Capital Market
1999. ISBN 3-7908-1193-9

M. Bonilla/T. Casasus/R. Sala (Eds.)
Financial Modelling
2000. ISBN 3-7908-1282-X

S. Sulzmaier
Consumer-Oriented Business Design
2001. ISBN 3-7908-1366-4

C. Zopounidis (Ed.)
New Trends in Banking Management
2002. ISBN 3-7908-1488-1

U. Dorndorf
Project Scheduling with Time Windows
2002. ISBN 3-7908-1516-0

B. Rapp/P. Jackson (Eds.)
Organisation and Work Beyond 2000
2003. ISBN 3-7908-1528-4

M. Grossmann
Entrepreneurship in Biotechnology
2003. ISBN 3-7908-0033-3

H. M. Arnold
Technology Shocks
2003. ISBN 3-7908-0051-1

T. Ihde
Dynamic Alliance Auctions
2004. ISBN 3-7908-0098-8

J. Windsperger/G. Cliquet/G. Hendrikse/M. Tuunanen (Eds.)
Economics and Management of Franchising Networks
2004. ISBN 3-7908-0202-6

K. Jennewein
Intellectual Property Management
2004. ISBN 3-7908-0280-8

M. J. Thannhuber
The Intelligent Enterprise
2005. ISBN 3-7908-1555-1

C. Clarke
Automotive Production Systems and Standardisation
2005. ISBN 3-7908-1578-0

M. Lütke Entrup
Advanced Planning in Fresh Food Industries
2005. ISBN 3-7908-1592-6

U. M. Löwer
Interorganisational Standards
2006. ISBN 3-7908-1653-1

G. Reepmeyer
Risk-sharing in the Pharmaceutical Industry
2006. ISBN 3-7908-1667-1

E. Kasper
Internal Research & Development Markets
2006. ISBN 3-7908-1728-7

L. Coleman
Why Managers and Companies Take Risks
2006. ISBN 3-7908-1695-7

M. A. Bader
Intellectual Property Management in R&D Collaborations
2006. ISBN 3-7908-1702-3

David L. Cahill

Customer Loyalty in Third Party Logistics Relationships

Findings from Studies in Germany and the USA

With 30 Figures and 114 Tables

Physica-Verlag
A Springer Company

Series Editors

Werner A. Müller
Martina Bihn

Author

Dr. David L. Cahill
Walther-von-der-Vogelweide-Str. 42
97074 Würzburg
Germany
logistics@cahill.de

ISSN 1431-1941
ISBN-10 3-7908-1903-4 Physica-Verlag Heidelberg New York
ISBN-13 978-3-7908-1903-8 Physica-Verlag Heidelberg New York

Physica-Verlag is a part of Springer Science+Business Media

springer.com

Typesetting: Camera ready by the author
Cover-design: Erich Kirchner, Heidelberg
Production: LE-TEX Jelonek, Schmidt & Vöckler GbR, Leipzig

SPIN 11865650 88/3100YL 5 4 3 2 1 0 Printed on acid-free paper

To my parents, Birgit and Stephen B. Cahill

Foreword

This dissertation by Cahill is embedded in a research stream at the Kuehne-Center for Logistics Management at WHU – Otto Beisheim School of Management – which explores customer loyalty towards logistics service providers. Underlying this is the work by Wallenburg (2004), which has been awarded the German Science Award for Logistics 2004 by the German logistics association BVL. Wallenburg's conceptual fundament is also guiding Cahill, whose work is empirically backed by two parallel large-scale surveys conducted in Germany and – shortly thereafter – in the USA. The latter survey was conducted in cooperation with The Ohio State University. The main characteristics of Cahill's dissertation have thus been outlined: (1) He replicates the study conducted by Wallenburg and additionally elaborates specific topics which have been identified as requiring further attention in research. And even though replications do not appear overly exciting prima facie, the scientific community acknowledges that they are indispensable and meaningful. (2) Cahill chooses an empirical research design and includes (3) intercultural management research. These three characteristics give Cahill's dissertation a very specific and incisive profile.

Cahill's work is grounded on a solid theoretical fundament, employing social exchange theory, equity theory, and commitment trust theory. As such, he includes theories which are prominent in customer loyalty research and intensively builds on Wallenburg's work. Even though this is reasonable and necessary, as he replicates Wallenburg's study, excluding those theories which did not contribute significant explanatory value in Wallenburg's work (e.g. transaction cost theory) demonstrates the autonomy of Cahill's deliberations. Embracing cultural theories additionally distinguishes Cahill's from Wallenburg's work. By including research by Hofstede, Hall & Hall, and Trompenaars & Hampden-Turner, Cahill selected three approaches predominantly found in the relevant literature. He identifies commonalities and differences between the approaches and consolidates the findings for use in his work, concluding that Germany and the USA are marked by high cultural similarity on the one hand, and some decisive differences on the other hand. Overall, Cahill builds a solid theoretical fundament for the empirical studies to be based upon.

Cahill's methodical considerations attest both a very high level of knowledge, as well as remarkable didactical skills. The cooperation with a research partner in the US ensured the quality of the American survey with regard to comprehensibility and consistency. The deployed statistical method suits the purposes of the research and is the state of the art for the research objectives. The careful preparation of the surveys ultimately surfaces in the samples' "technical data": sample size is sufficient in both countries, and even excellent in Germany. Also, response rates are convincing and in both samples, responses are representative of the included industries. In the end, the dual character of the study with parallel surveys in two countries sets Cahill's work apart from many others: comparative empirical research is – not only because of the considerable operational challenges – rare.

At this point, I will not divulge the results of the empirical study – I do not want to keep you from intensively studying this book. I will only reveal this: it is worth it!

Vallendar, June 2006

Prof. Dr. Jürgen Weber

Preface

It is a common misconception that writing a dissertation is a solitary endeavor. Rather, successfully researching a complex topic requires support from and interaction with a multitude of actors, as well as an environment that is liberal and stimulating. All this I have found at the Kuehne-Center for Logistics Management at WHU – Otto Beisheim School of Management – and my special gratitude goes to my doctoral advisor, Prof. Dr. Jürgen Weber. I would like to thank him for his wise guidance as well as the freedom bestowed upon me during the years working with him.

I would also like to thank my co-advisor, Prof. Dr. Lutz Kaufmann, who holds the Chair for International Management, for providing additional insights, especially with regard to the intercultural part of my dissertation.

Dr. Carl Marcus Wallenburg, Managing Director of the Kuehne-Center for Logistics Management, provided the fundament of this research with his dissertation "Kundenbindung in der Logistik" (Customer Loyalty in Logistics) published in 2004. I would like to thank Carl Marcus for providing the impulse to research third party logistics (3PL) relationships. Since my work is a continuation of his, I could not have wished for a better sparrings partner and critical counterpart.

From the very beginning, my work has been intimately interlinked with that of Dr. Jan M. Deepen. While my dissertation takes the service provider's view on 3PL relationships, Jan's takes the customer's view. In addition to sharing the survey data, Jan and I have been on the same time track in writing our dissertations and I am convinced that our work greatly benefited from our (at least daily) telephone conferences. For this, I would like to thank Jan, as much as for being a friend and an ally.

Research requires adequate funding. In this respect, I would like to thank the sponsors of my work: the Kuehne-Foundation, which finances a considerable portion of the research at the Kuehne-Center and my appreciation goes to Klaus-Michael Kühne, sole trustor of the foundation, and to its managing director Martin Willhaus. I would also like to thank Dr. Thomas Held, former CEO of Schenker AG, and Dirk Reich, Executive Vice President Contract Logistics of the Kuehne + Nagel Group, for specifically sponsoring the two surveys.

As I have pointed out above, the environment at the Kuehne-Center was critical for my dissertation's success. This is largely based on the great mix of researchers there and my thanks go to Dr. Andreas Gebhardt, Dr. Alexandra Matthes, Peter Voss, Wolfdieter Keppler, Serena Trelle, Peter Lukassen, Matthias Mahlendorf, Martina Bender, and Christian Busse. Also, work would have been a lot harder without our administrative assistants Beata Kobylarz, Fotini Noutsia, and Sonja Schmitt.

An important aspect of this dissertation is the comparison of Germany and the USA with regard to the formation of customer loyalty. These analyses are based on empirical data from both countries and I would like to thank Prof. A. Michael Knemeyer of The Ohio State University and Prof. Thomas J. Goldsby of the University of Kentucky. Tom and Mike facilitated data collection in the USA and became friends during our frequent visits.

Last but not least I would like to thank my family: my grandfather, Dr. Otto Lersch, for bestowing in me the motivation to acquire a PhD; Frank, for being there for me throughout the whole process, for looking after my culinary requirements, and for providing distraction when I needed it. Finally, my utmost gratitude goes to my parents. They have taught me to aim high and support me in whatever I pursue. Without them, I would not be where I am now, and to them, I dedicate this book.

Würzburg, July 2006

Dr. David L. Cahill

Contents

1 Introduction

1.1 Research motivation

A multitude of companies today has already identified the need to create a loyal customer base and acknowledges that maintaining existing customers and extending business with them is significantly less expensive than acquiring new customers (Stone, Woodcock, and Wilson 1996). Empirical proof of the proliferation of such customer loyalty efforts in the business world is e.g. provided in the form of loyalty programs, which many companies have installed during the past years (Hallberg 2004). By engaging in efforts aimed at creating customer loyalty, which in turn fosters financial success in monetary terms (Anderson, Fornell, and Lehmann 1994), firms react to increasing competitive challenges.

Within research, the investigation of customer loyalty gained importance when the classic marketing paradigm with its instrumental and transactional orientation proved unsuitable in the context of longer-term business relationships (Wallenburg 2004). Instead, the relationship marketing approach, which is specifically concerned with the study of relational exchanges (Morgan and Hunt 1994), gained importance within research, serving as a conceptual foundation for the majority of customer loyalty researchers.

The question of how loyalty develops has been subject to an abundance of research, leading to an expansive body of literature on loyalty determinants (Pritchard, Havitz, and Howard 1999). The extant literature exploring different factors and their constituent effects on loyalty, however, reveals a strong focus on consumer goods and industrial equipment settings, while industrial services have received relatively little attention so far. In addition, the majority of articles incorporates merely a few potential determinants and thus fails to draw a comprehensive picture of the mechanisms of customer loyalty formation.

Just like other businesses, logistics service providers (LSPs) are faced with increasing competitive pressure that urges them to concentrate not only on operational business processes, but also on an efficient and effective customer management. In the US alone, LSPs' revenues grew from

US-$ 31 billion in 1995 to US-$ 85 billion in 2004 and logistics outsourcing expenditures as a fraction of total logistics expenditures are at over 40% and expected to rise even further (Langley et al. 2005). One way to meet this challenge of rapid growth and expansion, according to Langley et al. (2005) is to focus on establishing, maintaining, and developing relationships with customers.

An often proposed driver of logistics outsourcing is the need to develop and maintain competitive advantage, which customers of LSPs intend to achieve through concentrating on core competencies and re-engineering (Sheffi 1990). Another important driver is the ongoing globalization, which several authors regard as the most important challenge that companies are facing (Cooper 1993). In this context, LSPs can play an important role as facilitators of global trade (Zhu, Lean, and Ying 2002). Along with globalization, however, companies that outsource logistics activities increasingly try to consolidate the number of LSPs they use globally (Persson and Virum 2001). Therefore, LSPs do not only have to devise sustainable growth strategies, but also have to develop intercultural management competencies (Huang, Rayner, and Zhuang 2003), a challenge hardly addressed in LSP management literature.

While intercultural management deals with the influence of culture on management styles in different countries, it is also arguable whether a one-best-way management paradigm is applicable even within national confines (Chow, Heaver, and Henriksson 1994). LSPs' customers are extremely diverse and similarly, relationships between LSPs and their customers can be expected to exhibit momentous differences (Knemeyer and Murphy 2005). As such, it is a crucial management issue for LSPs to design their customer loyalty efforts in a manner that accounts for both cultural context and different relationship characteristics.

1.2 Research goals

As outlined in the preceding section, LSPs are confronted with diverse management challenges that result from continuous growth, globalization, and customer diversity. The aim of the present study therefore is to identify determinants of customer loyalty in relationships between LSPs and their customers[1] by explicitly considering different characteristics and cultural contexts of such relationships. In this sense, the present research is positioned at the interface of marketing and logistics and is intended to

[1] Unless noted otherwise, within this study the word 'customer(s)' will always refer to customers of LSPs.

contribute not only to logistics research, but also to research in marketing, customer loyalty, and cultural studies.

In order to address the concept of customer loyalty, it is important to understand the mechanisms underlying loyalty in the logistics outsourcing context. For this reason, the starting point of the present research will be the study of Wallenburg (2004), who studied customer loyalty within relationships between LSPs and their customers. On this basis, factors that can be surmised to determine customer loyalty in such relationships will be proposed and interdependencies between these factors will be identified. The resulting comprehensive explanatory model of customer loyalty will not only provide insights into the constitution of customer loyalty, but will also serve as the basis for subsequent analyses.

As stated previously, a globalizing marketplace and the need of LSPs to render logistics services on an international scale requires intercultural management competencies. Before being able to apply such management techniques, though, a thorough understanding of cultural differences between different countries is necessary. The present study will therefore provide a starting point for such analyses by investigating cultural differences between two important markets for logistics outsourcing, the USA and Germany. Particular differences between Germany and the USA will be identified and applied to the previously devised customer loyalty model. As a result, differences between the two countries with respect to the formation of customer loyalty can be inferred.

Finally, this study will investigate in how far different relationship conditions influence the development of customer loyalty. For this purpose, important relationship characteristics will be identified and their moderating influences on the customer loyalty model will be examined. This will provide information on the robustness of the customer loyalty model versus relational contingencies and will suggest if it is necessary to differentiate customer loyalty efforts accordingly.

1.3 Structure

In order to achieve the research goals outlined in the preceding section, this study will link conceptual considerations with empirical analyses. For this reason, the framework of empirical logistics research proposed by Mentzer and Kahn (1995) will be applied to organize the research at hand.

Following this introductory chapter, chapter 2 presents the basic concepts of this study. For that, customer loyalty and logistics as the two relevant research fields will be introduced. In chapter 2.1, an overview of developments in customer loyalty research will be given, different customer loyalty concepts will be outlined, dimensions of customer loyalty will be discussed, and previous research on customer loyalty determinants will be presented. Then, chapter 2.2 deals with logistics, delineating recent developments in research, addressing issues pertaining to logistics outsourcing, and identifying different types of LSPs. Chapter 2.3 eventually links the two disciplines by stating particular research needs that lead to the research questions to be examined in this research.

Chapter 3 lays the theoretical fundament of the present research. In chapters 3.1, 3.2, and 3.3, social exchange theory, equity theory, and commitment trust theory are briefly introduced, as these are suitable for devising a customer loyalty model. Then, three cultural frameworks are presented in detail in chapter 3.4, from which cultural differences between the USA and Germany are derived.

Building on the concepts outlined in chapter 2 and the theories and frameworks introduced in chapter 3, chapter 4 provides the conceptual basis for the following empirical analyses. After brief conceptualizations of customer loyalty and its determinants in chapters 4.1 and 4.2, hypotheses are proposed in chapter 4.3 to formulate a comprehensive explanatory model of customer loyalty. After that, important relationship characteristics are proposed in chapter 4.4 and corresponding hypotheses are stated on the moderating effects these factors are surmised to have on the customer loyalty model. Finally, chapter 4.5 derives hypotheses on differences between Germany and the USA within the customer loyalty model.

Chapter 5 begins by identifying a methodology suitable for approaching the previously proposed hypotheses (5.1), before information about the two data-sets from Germany and the USA is given in chapter 5.2. Then, structural equation modeling as the selected methodology is introduced in chapter 5.3 and chapter 5.4 discusses criteria for assessing measurement and structural models. Chapter 5.5 rounds off the methodology chapter by outlining procedures to be applied to model assessment and potential model modification.

In chapter 6, measurement models are operationalized and tested against the empirical data from Germany and the USA. For each construct in the customer loyalty model individually, the conceptualization provided in chapter 4 is used to devise a measurement model, which is tested using the measures of model fit stated in chapter 5. After that, discriminant validity is assessed for the entire model, before relationship characteristics are operationalized and tested in chapter 6.3.

Chapter 7 unites insights from all preceding chapters by presenting results of the empirical analyses. While findings from Germany are given in chapter 7.1, 7.2 contains results from the USA. For each of the two countries separately, hypotheses regarding the customer loyalty model will be tested first, such that empirically validated customer loyalty models are provided for Germany and the USA. Then, moderating effects of relationship characteristics are discussed for Germany and the USA individually, before results are compared in chapter 7.3. Here, structural differences between the two validated customer loyalty models are discussed first, before the hypotheses on cultural differences between the two models are examined.

The study closes with chapter 8, which provides an overview of the main results of the research. In addition, important managerial implications are outlined and recommendations for further research are given.

2 Basic concepts

Having briefly discussed the goals of the present research in the preceding chapter, this chapter will first provide the notional basis of the research project by introducing the concepts of customer loyalty and logistics. While a wealth of literature exists in both areas, only the most important issues for carrying out this study will be discussed. The chapter will conclude by identifying research deficits in logistics and customer loyalty, from which a set of research questions will be derived to guide this study.

2.1 Customer loyalty

Obtaining a thorough understanding of customer loyalty is a prerequisite for the execution of the research at hand. For that, the development of customer loyalty research within the framework of relationship marketing will be presented first, before different customer loyalty concepts will be introduced. From these concepts, a definition of customer loyalty for use in this study will be derived, before both consequences and antecedents of customer loyalty will be portrayed.

2.1.1 Developments in customer loyalty research

Since the beginning of the 1990s, customer loyalty has gained importance both in relationship marketing research and in business. In business, this can be attributed to changing market- and competition-environments (Pritchard, Havitz, and Howard 1999, p. 333). Due to a shift from a sellers' to a buyers' market and because of an increasing degree of globalization, most industries find themselves confronted with new challenges. In a first phase, firms tried to face these challenges by focusing on their internal processes and organizational structures, trying to achieve cost reductions by concentrating on internal improvements. A second phase of external focus followed, where firms directed attention to their customers, trying to retain existing ones and to win over new ones (churning). Since "acquiring new customers is much more expensive than keeping them" (Stone, Woodcock, and Wilson 1996, p. 676), "loyal customers [...] are the bedrock of any business" (Caruana 2000, p. 811). According to Fornell (1992, p. 7) and Fournier and Yao (1997, p. 451), market shares have to be secured by retaining customers in markets facing low growth-rates and high competition. To Aaker (1996, p. 106), "a loyal customer base represents a barrier to entry, a basis for a price premium, time to respond to competitor innovations, and a bulwark against deleterious price competition." Baldinger and Rubinson (1997, p. 41) show that "loyalty is critical to brand volume, is highly correlated to market share, and can be used as the basis of predicting future market share; consequently, understanding loyalty appears critical to any meaningful analysis of marketing strategy."

In marketing research, two trends mark the development of customer loyalty. While individual transactions initially were in the center of marketing research, the focus shifted towards analyzing relationships. Grönroos (1994, p. 347) states that the 'traditional' marketing concept of the marketing mix with its '4 Ps'[1], developed in the middle of the last century, had been the established approach until the 1990s. This approach, however, focuses solely on transactions, a deficit tackled by the relationship marketing approach. At the core of it is the study of relationships between buyers and sellers of goods or services, in contrast to merely examining transactions (Andersen 2001, p. 167). An often cited and comprehensive definition of relationship marketing is provided by Morgan and Hunt (1994, p. 22): "Relationship marketing refers to all marketing activities directed toward establishing, developing, and maintaining successful relational exchanges." Therefore, the relationship marketing approach provides a suitable environment in which customer loyalty research can be nested.

While the development of relationship marketing began in the early 1970s, it was not until the late 1980s that works from the 'Nordic School of Services', mainly by Grönroos and Gummesson (e.g. Grönroos 1989, Grönroos 1994, Grönroos 1995, Grönroos 1997, Gummesson 1997), initiated a paradigm shift that geared marketing towards the creation, conservation, and extension of buyer-seller relationships. Although relationship marketing today is widely accepted among marketing researchers, its promoters do not postulate the replacement of the transactional approach, but rather juxtapose the two approaches. Grönroos (1995), for example, delineates a strategy continuum, in which different goods require different degrees of transaction- and relationship-based marketing strategies.

As a result of the focus on relationships in marketing research, customer loyalty gained importance within research. But even though the topic was

[1] The '4 Ps' stand for product, price, placement, and promotion.

first addressed in 1923, when Copeland investigated brand-repurchase, customer loyalty research only picked up momentum in the past twenty years. Since then, the following four streams can be identified (Wallenburg 2004, pp. 10-11):

(1) Examining relationships or customer loyalty itself. This is made up of those parts of relationship marketing that deal with the creation and development of relationships, i.e. the actual buying process, and of works striving to define customer loyalty and to measure it.
(2) Examining the effect of customer loyalty, especially the link between customer loyalty and corporate success.
(3) Examining determinants of customer loyalty. In addition to identifying determinants of customer loyalty, attention is paid to explaining how they influence each other and what their impact on customer loyalty is.
(4) Examining customer loyalty management. This field analyzes how different measures can increase customer loyalty and how these measures can be combined to form an efficient customer loyalty management.

Before determining which stream the present study can be associated with, however, it is important to create a clear understanding of different customer loyalty concepts prevalent in research. This will be accomplished in the following section.

2.1.2 Customer loyalty concepts

Reviewing research, it becomes obvious that the notion of customer loyalty is blurred. At its core, customer loyalty deals with relationships between suppliers and their customers and can be distinguished from other loyalty aspects, such as brand loyalty, which refer to a more abstract attachment, such as that towards a brand.

Within German customer loyalty literature, the notion of customer loyalty is even more faceted (Fassnacht and Daus 2004, p. 18). The German word for customer loyalty, 'Kundenbindung', encompasses both 'customer loyalty' and 'customer retention', wherefore Diller (1995, pp. 5-9) distinguishes an active, supplier-focused component and a passive, customer focused component of customer loyalty or 'Kundenbindung'.

In the supplier-focused perspective, customer loyalty is seen as a bundle of measures that aim at improving relationships with customers. The supplier is in the center of attention and the customer is only regarded as the factor at which success of customer loyalty becomes manifest. Here it becomes clear that this approach contains a conceptual deficit. It is the customer who eventually decides on whether customer loyalty management is successful or not, because all activities undertaken by a supplier can only be geared at influencing customers to be loyal. A customer-focused perspective therefore has to be added to evaluate the success of customer loyalty management.

Within the customer-focused perspective, customer loyalty is conceptualized taking into account customers' complex characteristics. These can either be approached as customers' directly observable actions and/or take into account their attitudes and intentions. Since customers' actions are directly influenced by their attitudes and intentions, it is obvious that these have to be scrutinized to understand and manage loyalty.

A third perspective identified by Diller (1995) is a synthesis of the former two approaches. The relationship-focused perspective directly examines the relationship between suppliers and customers. Accordingly, the objects of study in this perspective usually are buying behavior in retail contexts and long-term relationships marked by frequent interaction between suppliers and buyers in industrial contexts.

It is clear that the supplier-focused perspective with its instrumental approach is significantly different from the other two approaches. Distinguishing the customer- and the relationship-focused perspective, however, is difficult, because both focus on the customer. Eventually, even within the relationship-based perspective, a customer decides on the initiation, continuation, and development of a relationship. Therefore, researchers like Henning-Thurau, Gwinner, and Gremler (2000, p. 370) postulate to focus exclusively on the customer-based perspective. And in fact, the customer-focused perspective dominates current research on customer loyalty. As this approach serves as the basis of this study, details of the customer-focused perspective are given below.

Within the customer-focused perspective, two separate approaches have to be distinguished, the behavioristic and the neo-behavioristic approach. While the behavioristic approach dominated research until the 1970s, Jacoby and Kyner (1973) introduced the neo-behavioristic approach. Instead of exclusively considering observable behavior, as in behaviorism, neo-behaviorism accounts for customers' attitudes and intentions. The two approaches will be detailed in the following two sections. In addition, the object of customer loyalty requires some clarification, which will be provided in section 2.1.2.3.

2.1.2.1 Behavioristic customer loyalty concepts

Behavioristic concepts of customer loyalty have been at the core of early marketing research and focus on customers' observable behavior, as e.g. in purchasing behavior. Accordingly, customer loyalty is established, when customers demonstrate consistency in their choice of supplier or brand. Yim and Kannan (1999, p. 76) speak of "hard-core" loyalty, when one product alternative is exclusively repurchased and of "reinforcing" loyalty, when customers switch among brands but repeat-purchase one or more alternatives to a significant extent. Similarly, Neal (1999, p. 21) defines customer loyalty as "the proportion of times a purchaser chooses the same product or service in a specific category compared to the total number of purchases made by the purchaser in that category". Pegging customer loyalty to purchasing behavior, however, is very critical, as Cunningham (1956, pp. 121-128) pointed out very early: there can be a multitude of factors affecting purchasing behavior, such as product availability or special deals, which are not grasped by looking at purchases alone. A main deficit of the behavioristic approach thus is that it does not look at the drivers behind purchasing behavior.

Another disadvantage of behavioristic customer loyalty concepts is their ex-post approach, as pointed out by Wallenburg (2004, pp. 16-17). When loyalty is only expressed through purchases, information on customers' actual loyalty status in between purchases is not available. Consequently, decreasing loyalty is only recognized after it manifests itself through changed purchasing behavior. Only in relationships with frequent interaction can a supplier integrate further aspects, such as complaints, into customer loyalty management. The reason, why behavioristic concepts may still be valuable, is because the measurement of customer loyalty in this approach does not necessitate involvement by the customer. The assessment of attitudes and intentions would always imply customers' cooperation through participation in surveys. By simply recording purchases, e.g. through delivery records in the industrial context or customer cards in a consumer context, the assessment of customer loyalty poses little difficulty. Particularly in areas, where most purchases can be easily ascribed to individual customers, as is the case with mail-ordering or book-stores on the internet, the behavioristic approach is useful for identifying different customer groups and their characteristics. Such firms, however, can only assess purchases of their own products, while purchases of competing products go unnoticed. Firms can therefore neither draw conclusions about relative changes of purchasing behaviors, nor evaluate their comparative market position.

2.1.2.2 Neo-behavioristic customer loyalty concepts

Too, Souchon, and Thirkell (2001, p. 292) criticize that "these narrow technical definitions do not adequately capture the richness and depth of the loyalty construct implicit in a relational framework." Consequently, neo-behavioristic customer loyalty concepts start at the shortcomings of the behavioristic approach by examining the causes of loyalty. As early as 1969, Day concluded that "loyalty should be evaluated with both attitudinal and behavioral criteria" (Day 1969, p. 30), because otherwise accidental repeat-purchases, merely resulting from situational exigencies, would be regarded as indicators of loyalty.

There is no agreement, however, on the question, whether attitudes are part of customer loyalty or merely an antecedent of it. Some authors, e.g. Dick and Basu (1994), propose that only positive attitude can lead to 'true' customer loyalty. If attitude then is a necessary prerequisite of customer loyalty, some drivers of loyalty cannot be explained. Transaction cost theory, for instance, provides the concept of asset specificity (e.g. Williamson 1985, p. 2 and Williamson 1990, p. 142). Relationship-specific investments create economic switching barriers and therefore increase customer loyalty. According to Dick and Basu (1994), however, the mere repeat purchase of goods or services for reasons of economic constraints would not qualify as loyalty, as positive attitudes are not involved. In order to avoid the outlined problem, it is useful to abstain from defining positive attitude to be a necessary antecedent of loyalty. Instead, researchers usually consider intentions and observable behavior to be the constituting elements of customer loyalty. Referring to intentions instead of attitudes has the advantage that intentions are value-free, and therefore also allow for 'involuntary' customer loyalty that may result from economic constraints (Bendapudi and Berry 1997, p. 28, Kumar, Bohling, and Ladda 2003, p. 669).

2.1.2.3 Dimensions of customer loyalty

For some years, researchers have been stipulating the multi-dimensionality of customer loyalty. According to Meyer and Oevermann (1995), purchasing (repeat purchases of the same goods or services and additional purchases of different goods or services) and referrals form customer loyalty. While purchasing is totally established in customer loyalty research, the inclusion of referrals, or word-of-mouth marketing[2], accounts for the fact that referrals are the highest expression of positive attitude towards a sup-

[2] See Wilson and Peterson 1989 and Wilson 1994 for details on word-of-mouth marketing.

plier. At the same time, they distinguish between behavior and intentions, therefore following the neo-behavioristic school of thought.

This multi-dimensional conceptualization, as depicted in Figure 2-1, however, is not frequently found in empirical studies. Until today, a restricted approach, oftentimes only including purchases, dominates empirical research on customer loyalty (e.g. Bloemer and de Ruyter 1999, Gronholdt, Martensen, and Kristensen 2000, Sivadas and Baker-Prewitt 2000, Chaudhuri and Holbrook 2001, Crutchfield 2001, De Wulf, Odekerken-Schröder, and Iacobucci 2001, Lee, Lee, and Feick 2001, Olsen 2002, Sirdeshmukh, Singh, and Sabol 2002).

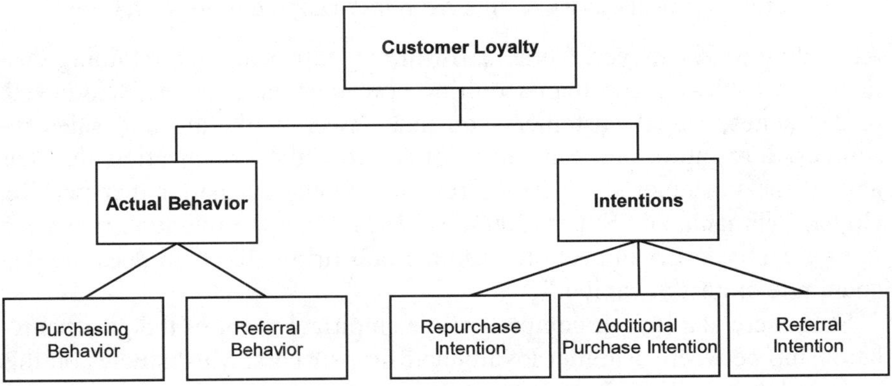

Fig. 2-1. Conceptualization of customer loyalty[3]

Analyses of customer loyalty research show that most works do not include both behavior and intentions. In empirical research on long-term relationships, such as that on relationships between logistics service providers and their customers, actual behavior is usually omitted, tending to the impossibility of observing purchasing behavior. Instead, such research in most cases focuses on intentions, assuming that intentions strongly influence behavior.

2.1.3 Effects of customer loyalty

Customer loyalty cannot be an end in itself and research is only justified if it has substantial impact on firms' success. An examination of literature reveals that customer loyalty is usually assumed to have a positive influ-

[3] Adapted from Meyer and Oevermann 1995, p. 1342.

ence on success. Reichheld, Markey, and Hopton (2000, p. 135) state the following three effects that customer loyalty induces:

(1) "Revenues and market share grow as the best customers are swept into the company's book of business, building repeat sales and referrals.
(2) Costs shrink as the expense of acquiring and serving new customer and replacing old ones declines.
(3) Employee retention increases because job pride and job satisfaction increase, in turn creating a loop that reinforces customer retention through familiarity and better service to the customers. Increased productivity results from increasing employee tenure."

According to Knemeyer, Corsi, and Murphy (2003, p. 84), retaining customers is far less costly than acquiring new ones, because, as Neal (1999, p. 21) states, "loyal customers consume fewer marketing and sales resources, buy more, and buy more often from the organization that has gained the customer's loyalty." From a financial valuation perspective, Gupta, Lehmann, and Stuart (2004, p. 17) find that the retention rate has a "significantly larger impact on customer and firm value than does the discount rate or cost of capital."

Still, there is a lack of comprehensive empirical proof of the positive relationship between customer loyalty and firm success. Must studies on this topic solely examine the effect on revenues and rely on case studies or small-scale surveys. Even the often cited works by Reichheld and Sasser, e.g. Reichheld and Sasser (1990), are merely a collection of case studies which cannot be the basis for generalized conclusions. One of the few large-scale, empirical studies examining the effects of customer loyalty was conducted by Kalwani and Narayandas (1995), showing that companies with higher customer loyalty exhibit higher revenues, higher returns on investment, and a higher profitability of innovations.

Apart from empirical evidence, however, it is reasonable to argue that customer loyalty has effects both on costs and on revenues. Since customers are the main source of revenue, any company has to gear its efforts towards the acquisition of new customers and the retention of existing ones. However, companies have to bear in mind that customer loyalty also leads to costs arising from measures aimed at increasing customer loyalty, but also from product improvements that increase customer satisfaction and eventually customer loyalty (Reinartz, Thomas, and Kumar 2005). Therefore, customer loyalty can only be assumed to have a positive impact on success, if its utility is in proportion to its cost (Wallenburg 2004, p. 27).

2.1.4 Determinants of customer loyalty

In order to be able to gear marketing activities towards the creation of customer loyalty, its determinants and their precise effects have to be known. Accordingly, many researchers have investigated this topic. In order to gain an overview of the determinants identified in these works, they can be structured in three dimensions (Fassnacht and Daus 2004, p. 5-7):

(1) Company-related determinants refer to the supplier itself or to the goods or services offered. It is a prerequisite for the existence of customer loyalty that the offered goods or services create utility for the customer and that they are available. In this respect, an assessment is usually performed by examining quality. In order to evaluate the price-performance ratio, customers will pay attention to prices. Customer loyalty will also be influenced by the reputation a company has and ultimately by customer loyalty programs offered.

(2) Relationship-related determinants play a significant role in long-term relationships. Factors regarding the interaction between supplier and customer, such as relationship quality, previous experiences, and trust are important. Commitment, which provides evidence of emotional closeness and moral or normative feelings of obligation, takes a central role in relationships. Specificity and dependence can lead to economic, psychological and social switching barriers.

(3) Customer-related determinants are mainly influenced by customers' characteristics. In this respect, affect and involvement, and consequently also the importance of the good or service to the customer, are important.

In addition to the above delineated areas, the effects of the market environment and competition are researched, as is the link between satisfaction and loyalty, which plays an important role in the research of customer loyalty and is often placed in one of the three dimensions. However, as most other determinants influence satisfaction, it cannot be clearly separated and should therefore be listed as a distinct category.

Table 2-1. Determinants of customer loyalty in empirical research[4]

	Prices and Conditions	Product and Service Quality	Product Line, Availability, Accessibility	Image and Reputation	Customer Loyalty Programs	Intensity and Quality of Business Relationship	Own Experiences and Recommendations from Others	Trust	Commitment	Economic Switching Barriers	Psychological and Social Barriers to Switching	Sociodemographic Customer Attributes	Affect	Involvement	Importance of the Product for the Customer	Pattern of Customers' Behavior	Expectations	Utility from Consumption/Business Relationship	Attractiveness of Competitios' Offers/Incentives to Change	Customer Satisfaction
Bitner (1990)		●																		●
Sriram and Mummalaneni (1990)										●									●	
Bloemer and Lemmink (1992)																				●
Cronin Jr. and Taylor (1992)		●																		●
Fornell (1992)		●								●	●						●			●
Anderson and Sullivan (1993)		●															●			●
Biong (1993)	●	●																		●
Boulding et al. (1993)		●															●			
Ping (1993)										●									●	●
Rust and Zahorik (1993)	●	●	●			●		●												
Selnes (1993)		●		●																●
Ganesan (1994)				●			●	●		●										●
Taylor and Baker (1994)		●																		●
Bloemer and Kasper (1995)																●				●
Keaveney (1995)	●	●	●			●									●				●	●
Fornell et al. (1996)		●															●	●		●
Hallowell (1996)																				●
Zeithaml, Berry, and Parasuraman (1996)																				●
Macintosh and Lockshin (1997)						●		●												●
Oliver, Rust, and Varki (1997)													●				●			●
Sharp and Sharp (1997)					●															
Bolton (1998)							●													●

●= in the empirical study listed in the row, the determinant listed in the column had an effect on customer loyalty.

[4] Adapted from Fassnacht and Dauer 2004, pp. 8-9 and extended for 2004-2005.

Table 2-1. Determinants of customer loyalty in empirical research (cont.)

	Prices and Conditions	Product and Service Quality	Product Line, Availability, Accessibility	Image and Reputation	Customer Loyalty Programs	Intensity and Quality of Business Relationship	Own Experiences and Recommendations from Others	Trust	Commitment	Economic Switching Barriers	Psychological and Social Barriers to Switching	Sociodemographic Customer Attributes	Affect	Involvement	Importance of the Product for the Customer	Pattern of Customers' Behavior	Expectations	Utility from Consumption/Business Relationship	Attractiveness of Competitios' Offers/Incentives to Change	Customer Satisfaction
de Ruyter, Wetzels, and Bloemer (1998)		●																		
Mittal, Ross, and Baldasare (1998)																	●			●
Bloemer, de Ruyter, and Wetzels (1999)		●																		
Bloemer and de Ruyter (1999)													●							●
de Ruyter and Bloemer (1999)																				
Garbarino and Johnson (1999)													●					●		●
Mittal, Kumar, and Tsiros (1999)																				●
Pritchard, Havitz, and Howard (1999)									●		●			●						
Reynolds and Beatty (1999)															●			●		●
Bolton, Kannan, and Bramlett (2000)	●	●			●		●												●	●
Gronholdt, Martensen, and Kristensen (2000)		●		●													●	●		●
Chaudhuri and Holbrook (2001)								●					●					●		
De Wulf, Odekerken-Schröder, and Iacobucci (2001)					●	●		●	●											●
Mittal and Kamakura (2001)												●								●
Too, Souchon, and Thirkell (2001)			●		●			●	●											
Olsen (2002)		●																		●
Roehm, Pullins, and Roehm Jr. (2002)					●															
Sirdeshmukh, Singh, and Sabol (2002)								●										●		
Hellier et al. (2003)																				●
Homburg, Giering, and Menon (2003)						●	●	●		●	●									●
Ranaweera and Prabhu (2003)								●		●										●

Table 2-1. Determinants of customer loyalty in empirical research (cont.)

	Prices and Conditions	Product and Service Quality	Product Line, Availability, Accessibility	Image and Reputation	Customer Loyalty Programs	Intensity and Quality of Business Relationship	Own Experiences and Recommendations from Others	Trust	Commitment	Economic Switching Barriers	Psychological and Social Barriers to Switching	Sociodemographic Customer Attributes	Affect	Involvement	Importance of the Product for the Customer	Pattern of Customers' Behavior	Expectations	Utility from Consumption/Business Relationship	Attractiveness of Competitios' Offers/Incentives to Change	Customer Satisfaction
Shankar, Smith, and Rangaswamy (2003)																				•
Verhoef (2003)					•				•											
Yen and Gwinner (2003)								•										•		
Yi and Jeon (2003)					•															
Bansal, Irving, and Taylor (2004)									•											
Beerli, Martín, and Quintana (2004)										•										•
Guenzi and Pelloni (2004)						•														•
Harris and Goode (2004)								•										•		•
Hennig-Thurau (2004)						•			•											•
Lam et al. (2004)										•								•		•
Lewis (2004)					•															
Ngobo (2004)		•								•								•		•
Venetis and Ghauri (2004)		•							•											
Wallace, Giese, and Johnson (2004)																				•
Agustin and Singh (2005)								•										•		
Algesheimer, Dholakia, and Herrmann (2005)						•														
Bell, Auh, and Smalley (2005)		•																		
Borle et al. (2005)			•																	
Fullerton (2005)		•							•											
Gustafsson, Johnson, and Roos (2005)									•											•
Homburg and Fürst (2005)																				•
Jiang and Rosenbloom (2005)	•																			•
Lin and Ding (2005)						•		•												•

The above presented deliberations show that there is a multitude of potentially relevant factors determining customer loyalty. Table 2-1 therefore provides an overview of empirical studies concerned with the analysis of customer loyalty determinants, which is based on a meta-analysis conducted by Fassnacht and Daus (2004)[5]. It is clear, that the satisfaction-loyalty link has enjoyed the greatest attention in those works and that the loyalty effects of product and service quality are another important topic. Determinants that were researched by a considerable amount of studies deal with customer loyalty programs, trust, expectations, and utility. Very little attention so far has been ascribed to psychological and social barriers to switching, sociodemographic customer attributes, involvement, importance of products for customers, and customers' behavioral patterns. This leads to two conclusions: First, general agreement on which determinants to include in customer loyalty models has not been reached as yet. Secondly, most works only examine a very limited number of determinants, and a holistic approach to the explanation of customer loyalty is rarely provided (Ranaweera and Neely 2003, p. 230).

2.2 Logistics

After having presented the necessary fundamentals of customer loyalty in the preceding section, this section will elaborate on the reference point of the research project, i.e. logistics. For that, the first section will present different views on logistics and lay the notional basis for the understanding of logistics embraced in this study. As relationships between logistics service providers and their customers are to be analyzed, the logistics outsourcing phenomenon will be detailed, before a taxonomy of logistics service providers is provided.

2.2.1 The nature of logistics

Logistics is an established discipline both in theory and in practice, yet there is no agreement on a universal definition of logistics. One widely accepted view stems from the Council of Supply Chain Management Professionals (CSCMP). In their latest definition (CSCMP 2005), logistics man-

[5] Since the overview provided by Fassnacht and Daus 2004 only covers articles published until 2003, Table 2-1 was extended to include English-language articles published in 2004 and 2005 in the journals selected by Fassnacht and Daus 2004.

agement is seen as "[…] that part of Supply Chain Management that plans, implements, and controls the efficient, effective forward and reverse flow and storage of goods, services and related information between the point of origin and the point of consumption in order to meet customers' requirements."[6] This view is clearly marked by the economic objectives of effectiveness and efficiency and regards logistics as part of supply chain management.

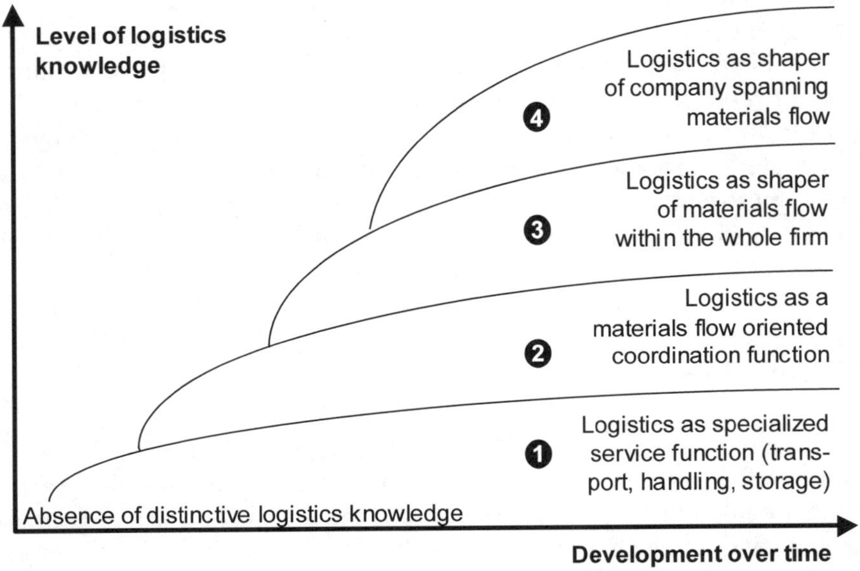

Fig. 2-2. Logistics development stages[7]

Other definitions, e.g. that of Weber and Kummer (1998, pp. 7-28), are broader and refer to logistics as a flow-oriented design of all value-creation processes. At the base of their understanding is the empirically backed assumption that there are different stages of maturity in logistics. While there

[6] CSCMP 2005 defines supply chain management as "[…] the planning and management of all activities involved in sourcing and procurement, conversion, and all Logistics Management activities. Importantly, it also includes coordination and collaboration with channel partners, which can be suppliers, intermediaries, third-party service providers, and customers. In essence, Supply Chain Management integrates supply and demand management within and across companies." See Larson and Halldorsson 2004 for more details on supply chain management.

[7] Adapted from Weber 2002, pp. 9-10.

are several researchers that also argue for stage models of logistics development, there is no unity as to how many stages such a model has. Some researchers, such as Bowersox and Daugherty (1987) and Klaus (1999) argue for three stages and Cavinato (1999) even sees five stages. The model that will be briefly examined in this section stems from Weber (2002), who identifies four maturity stages. These stages are determined by the level of logistics knowledge present in a firm and require path-dependent development from the lowest to the highest level of logistics knowledge:

In the first maturity stage, logistics is a specialized service function that provides transportation, handling, and storage services that are necessary for an efficient flow of materials and goods. Weber (2002, p. 7) states that this stage of logistics maturity was first observable in the 1950s, when markets changed from suppliers' to buyers' markets, requiring improved materials flows to succeed in the changed market environment. This is in line with Bowersox and Daugherty (1987, p. 47), who confirm that advanced logistical organizations barely existed at that time. Effects of this paradigm-change were two-fold. On the one hand, logistics optimization was achieved through process improvements and advances in forecasting and planning techniques. On the other hand, organizational changes took place, as many companies institutionalized their logistics functions in dedicated departments and therefore created a specialized service function that was separated from other functions such as procurement or manufacturing.

The second stage of logistics maturity was initiated by contextual changes in the economic environment. According to Ballou (2003, pp. 1-30), the oil crisis in the early 1970s and the following inflation and stagnation of the economy made companies seek ways to create competitive advantages. In addition, advances in technology facilitated communication and networks between different departments in companies, as Bowersox, Closs, and Helferich (1986, p. 14) state. As a result, firms realized that efficiency could be increased by improving the coordination of materials flows from inbound streams in procurement through the value creation process in manufacturing to outbound streams in distribution. Through this focus on the integration of different functions, cost and performance benefits were achieved. Weber (2002, p. 11) provides coordination of lot-sizes and just-in-time supply and production as examples of these benefits.

The following third stage of logistics maturity was necessitated by further changes of the market environment. Simon (1988) points out that the intensity of competition increased, an effect attributable to over-capacities, a world-wide convergence of product quality, shortened product-life-cycles, and a strengthened focus on customer demands. Wallenburg (2004, p. 42) concludes, that the required simultaneous focus on differentiation and cost leadership was not feasible under the prevalent functional structures and rigid organizational systems. Instead, a process-oriented intra-company value chain that reduces complexity was needed to succeed in the changing market environment. Hence, logistics evolved from a mere functional concept to a management concept, or, as Bowersox and Daugherty (1987, p. 47) put it, from operational to strategic orientation.

The fourth and, for the time being, last maturity stage of logistics development is again driven by market pressures. When companies realized that optimization potentials within the organization had been largely exploited, the focus of attention was furthered to include upstream and downstream partners in the supply chain (Bolumole 2001, p. 88). In this sense, the view of logistics as a management concept from the third maturity stage is extended over company boundaries and can be labeled supply chain management (SCM). According to a study conducted by Larson and Halldorsson (2004), SCM in this sense is a sub-group of logistics.

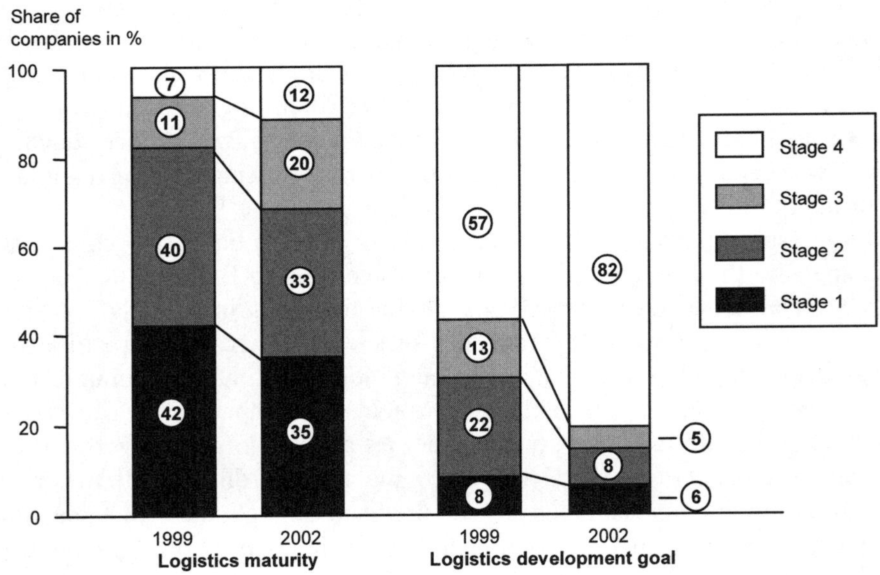

Fig. 2-3. Maturity stages and development goals of logistics[8]

This four-stage model of logistics development as proposed by Weber and Kummer (1998) and refined by Weber (2002) underwent empirical examination by Weber and Dehler (1999) and by Wallenburg (2004). In these studies conducted in Germany, evidence for the existence of the four

[8] Adapted from Wallenburg 2004, p. 45.

stage model was found. The results from the two studies are given in Figure 2-3 and indicate that logistics has actually advanced since the 1999 study. Still, nearly 70% of respondents are on the first two stages of logistics maturity, because the functional view of logistics still prevails in their companies. In addition, the two studies also determined the development goals that respondents had for their logistics. It shows that the majority of respondents want to develop their logistics towards the highest maturity stage, i.e. SCM.

As shown in Figure 2-3, logistics is on a development path towards higher stages in the logistics maturity model. This quest, however, is only reasonable, if higher levels of logistics development lead to corporate success. Therefore, Dehler (2001, pp. 220-226) scrutinized the link between flow-orientation and success in an empirical study in Germany. He found that flow-orientation increases logistics success by lowering logistics cost and improving logistics service, which is also supported by findings from Engelbrecht (2003, p. 64). Logistics success, in turn, has a positive effect on corporate success. These results were recently confirmed by Engelbrecht (2004, pp. 251-255) and Deepen (2006).

This underlines the outstanding importance of logistics for companies, which can realize significant economic benefits by allocating management capacities to the improvement of logistics and the creation of flow-oriented organizations. A way, in which many companies presently try to accomplish this is by outsourcing parts or all of their logistics activities to third parties. Details on logistics outsourcing will therefore be provided in the next sections.

2.2.2 Logistics outsourcing

When a company considers the organization of its logistics functions, it is faced with three basic options. According to Razzaque and Cheng (1998, p. 89), a company can provide the function in-house by making the service, own logistics subsidiaries through setting up or buying a logistics firm, and/or outsource the function and buy the service. In the following section, research on the last option, the outsourcing of the logistics function, will be discussed, as outsourcing relationships are at the core of the research at hand.

2.2.2.1 Definitions and extent of logistics outsourcing

Before emerging into a detailed discussion of logistics outsourcing, several definitions of the logistics outsourcing phenomenon will be presented.

First, notional clarification has to be provided regarding the use of the terms 'logistics outsourcing', 'contract logistics', and 'third-party logistics'. Although textually distinct in the German language, Knemeyer and Murphy (2004, p. 35) maintain that the three terms are used synonymously, which is also manifested in the following definitions:

Lieb, Millen, and van Wassenhove (1993, p. 35) choose a very broad approach, by saying that companies that outsource "employ an outside company to perform some or all of the firm's logistics activities. The arrangement may be narrow in scope (for example, limited to warehouse services only), or broad, encompassing the entire supply chain." In contrast to this, Bradley (1994a, p. 56A3) more narrowly defines logistics outsourcing as a relationship in which the service provider offers "at least two services that are bundled and combined, with a single point of accountability using distinct information systems that are dedicated to and integral to the logistics process". Newer definitions are a lot broader again, such as those of Lambert, Emmelhainz, and Gardner (1999, p. 165) who state that logistics outsourcing is "the use of a third-party provider for all or part of an organization's logistics operations" or of Rabinovich et al. (1999, p. 353), who define logistics outsourcing as "long and short-term contracts or alliances between manufacturing and service firms and third party logistics providers". In contrast to the aforementioned definitions, Africk and Calkins (1994, p. 49) take a relational position towards logistics outsourcing by defining "contract logistics as the outsourcing of all or part of a company's logistics function. Relative to basic services, contract logistics offerings are more complex, encompass a broader number of functions, and are characterized by longer-term, more mutually beneficial relationships." Hence, this definition appears most suitable in the context of this study focusing on customer loyalty in the context of relationships between logistics service providers and their customers and will therefore serve as the definitional base of the notion of logistics outsourcing.

While logistics outsourcing has been widely established in Europe for a long time (Bardi and Tracey 1991), Barks (1994, p. 36) states that only $10 billion were spent on logistics outsourcing in the U.S. in 1993. While this number was up to $46 billion by 1999, as Cooke (2000, p. 69) says, it was approaching $80 billion in 2004 (Gecker 2004). In an empirical study conducted in 2000, Lieb and Miller (2002, p. 7) find that surveyed Fortune 500 companies in 1999 and 2000 spent 19% of their annual logistics operating budgets on logistics outsourcing, but expected to spend 31% three years later. According to Langley et al. (2005, p. 12), logistics outsourcing expenditures as a fraction of total logistics expenditures are already at over 40% and expected to rise further. The following sections will shed light on the reasons underlying the rising importance of logistics outsourcing in business practice.

2.2.2.2 Drivers of logistics outsourcing

Although there is no doubt about the rising importance of logistics outsourcing in business practice, different drivers of this development have been identified in research. According to Hill (1994, pp. 28-30), Lieb (1992, p. 29), and Sheffi (1990), the main goals of outsourcing logistics functions to third parties are the need to develop competitive advantage through a growing emphasis on providing good customer service effectively and efficiently and the strategic value of focusing on core businesses and re-engineering. Gaining competitive advantage, however, has become more difficult in a globalizing world. As international growth and sourcing demand improved logistics processes, as proposed by McCabe (1990), Whybark (1990), Bovet (1991), Cooper (1993), and Fawcett, Birou, and Taylor (1993), many authors, including Trunick (1989), Foster and Muller (1990), Sheffi (1990), Byrne (1993), Rao, Young, and Novick (1993), and Sum and Teo (1999) regard globalization as the most important driver of logistics outsourcing activities. Another driver of the logistics outsourcing trend is seen in sophisticated logistical concepts such as just-in-time (JIT). Since these concepts demand special expertise and added resources, Trunick (1989), Goldberg (1990), and Sheffi (1990) suggest that users of these concepts seek support from outside their organizations. Trunick (1989) and Lewis and Talalayevsky (2000) see another driver of logistics outsourcing in the emergence of new technologies that would be "time consuming and expensive to develop and implement [...] in-house" (Razzaque and Cheng 1998, p. 91).

In order to provide further insights into why businesses actually do engage in logistics outsourcing projects, the motivation to outsource was included in the survey that forms the empirical basis of this study (see chapter 5 for further details on the survey). Participants of the study were asked to indicate the major reasons for outsourcing parts of their logistics activities on a Likert-scale of 1 (does not apply) to 7 (fully applies). Multiple indications were allowed. The average results for the eleven factors included in the survey are presented in Figure 2-4 for the German data-set (see Figure 7-5 for results from the US) and are largely in line with the results obtained by Engelbrecht (2004, pp. 241-243) from his 2002 study. The current data supports the previous finding that financial considerations, i.e. the reduction of logistics costs, the variabilization of fixed cost, and the reduction of capital employed, are the most important reasons to outsource, while performance-related motives only play a subordinate role. While this shows, why companies actually engage in logistics outsourcing, the following sections will point out advantages and disadvantages associated with logistics outsourcing.

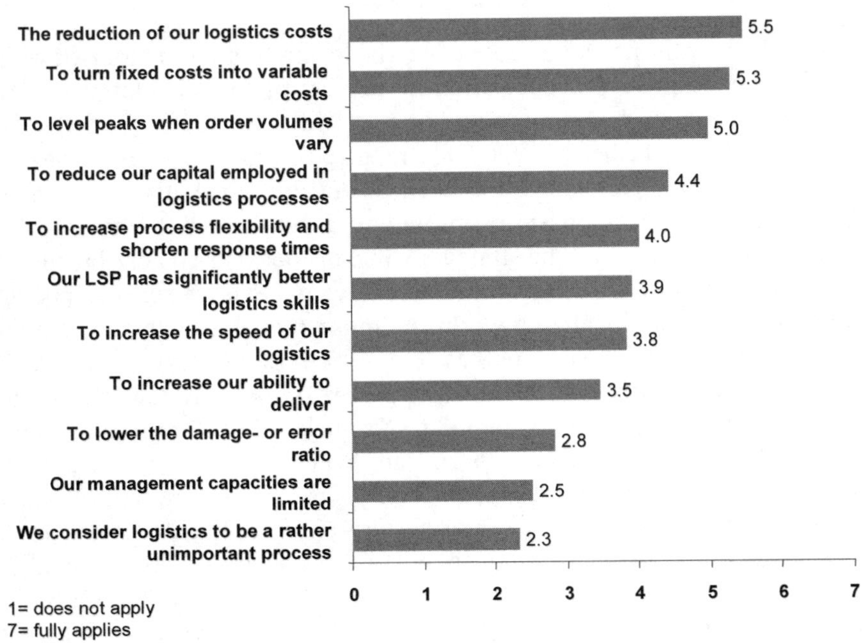

Fig. 2-4. Motivation to outsource in 2004[9] (Germany)

2.2.2.3 Advantages and disadvantages of logistics outsourcing

When outsourcing logistics functions, companies can enjoy many advantages but also have to be aware of the associated problems. The advantages of logistics outsourcing appear on two sides – there can be cost reductions and performance improvements (Skjoett-Larsen 2000, p. 112), which is also reflected in the actual outsourcing motivation as depicted above.

Most authors agree about the cost reduction potentials of logistics outsourcing (e.g. Lieb 1992, p. 29, Candler 1994, p. 46, Panayides 2004, p. 1), which can be achieved mainly in the following areas:

- Reduction of capital investments into facilities, equipment, and information technology (e.g. Foster and Muller 1990, Richardson 1992,

[9] Results from own study, see chapter 5.

Richardson 1995, Fantasia 1993, Goldberg 1990, Lacity, Willcocks, and Feeny 1995, Sheffi 1990, Trunick 1992).
- Less manpower needed (Foster and Muller 1990, Richardson 1992, Richardson 1995), which is especially relevant in markets with high wages and/or restrictive labor laws, such as Germany.

These cost reductions are facilitated by specific advantages that LSPs enjoy and which they can pass on to their customers, because

- they are more efficient and have lower production costs, resulting from the fact that logistics is their core competency (Bretzke 1993, p. 38);
- they can realize economies of scale and scope (Schäfer-Kunz and Tewald 1998, p. 30-36);
- they can equal out demand peaks (such as seasonal differences of volumes) by diversifying their customer portfolio (Deepen 2003, p. 130).

In addition, outsourcing allows companies to expose and eliminate inefficiencies that were unobservable when the logistics function was performed in-house (Wallenburg 2004, pp. 46-47) and according to Bradley (1994b) and Richardson (1993), fixed costs can be converted to variable costs.

But, as Heskett (1977, p. 85) states, "management cannot measure the importance of logistics in terms of cost alone." Rather, logistics outsourcing can increase performance, as companies can utilize expertise, technology, and infrastructure of LSPs (Browne and Allen 2001, pp. 259-260). Consequently, companies can benefit from higher quality, better service, optimized asset use, and increased flexibility (La Londe and Maltz 1992, p. 3), as well as faster transit times, less damage, and improved on-time delivery (Richardson 1990, Richardson 1995). The additional flexibility also makes companies more responsive to changes in marketing, manufacturing, and distribution (Browne and Allen 2001, pp. 259-260, Wallenburg 2004, p. 47). Furthermore, the core competency debate in the recent years suggests that companies focus their efforts on creating sustainable competitive advantage. In this context, Wallenburg (2004, p. 47) states that outsourcing can reduce the complexity of companies' business processes, thus facilitating competitive advantage.

Still, many authors point out problems that are created by outsourcing logistics functions. The main reason indicated by most authors is the loss of control over quality and service, efficiency, and price that ultimately lead to dependence (e.g. Wentworth 2003, pp. 57-58, Lynch, Imada, and Bookbinder 1994, p. 103). In addition to pointing out transactional risks, such as bounded rationality, opportunism, small numbers bargaining, and information impactedness, McIvor (2000) points out that many potential benefits from logistics outsourcing are often not leveraged, because of a short-term focus on cost reductions.

Consequently, companies intending to outsource logistics functions have to pay close attention to the associated risks and benefits (Razzaque and Cheng 1998, p. 96). As the preceding sections have shown, LSPs are enablers of logistics outsourcing and a typology of LSPs will therefore be provided in the following section.

2.2.3 Logistics service providers

If a company employs an outside provider to perform some or all of its logistics activities, this outside provider is termed a logistics service provider (Bhatnagar, Sohal, and Millen 1999, p. 569). These LSPs were originally only offering a very narrow spectrum of services, mainly consisting of transportation or warehousing services. In order to react to changing demands from their customers, as pointed out in the preceding sections, LSPs have started to offer integrated service portfolios that include a multitude of different services. According to Wallenburg (2004, pp. 50-51), five types of LSPs can currently be distinguished: carriers, couriers & express & parcel/postal (CEP), freight forwarders, third-party LSPs (3PLs), and fourth-party LSPs (4PLs), which will be briefly described below.

Carriers are firms that own assets for transportation purposes. These assets are usually confined to either road, sea, air, or rail transportation. In some cases, however, carriers own several of the aforementioned transportation assets. As carriers receive their orders either directly from shippers or from brokers, the management of capacity and load-factor optimization are at the core of the carrier business.

CEP providers are also transportation providers, but with a focus on the distribution of small units to any destination, often with time-critical shipments. As opposed to carriers, CEP providers send parcels or mail though their networks, employing different transportation vehicles and incorporating handling at hubs. In this way, they can virtually supply door-to-door services anywhere in the world, without having to rely on any third parties. Still, carriers and brokers are sometimes employed to reduce capital intensity. Core competencies of CEP providers are the operation and management of highly complex distribution networks.

Freight forwarders are intermediaries between suppliers and buyers of transportation services. In most cases, brokers carry out additional services, such as transportation planning and management, including the associated provision of information systems. In some cases, brokers also dispose of their own transportation assets. Bundling and coordination of demand for transportation services are the core competencies of brokers.

In contrast to the aforementioned LSPs, 3PLs or contract LSPs[10] offer their customers logistics solutions that often include management capacities. These solutions most commonly include several services, such as warehousing and pick/pack operations that are carried out on a longer-term contractual basis. Sometimes, 3PLs provide even more customized services that integrate into their customers' value chains, such as fleet management, order handling, complaints management, call centers, or assembly services.

A recent addition to the list of types of LSPs comes in the form of 4PLs. While there is no universal agreement on the definition of the 4PL, the current state of the debate establishes 4PLs as suppliers of complete logistics systems, without carrying out the services themselves. Instead, 4PLs subcontract all operations from other LSPs and confine themselves to the management of logistics systems. Thus, they do not dispose of any logistical assets and are a neutral intermediary between businesses and asset-based LSPs.

As 3PLs form the point of reference for logistics outsourcing as dealt with in this study, the existing notional ambiguities in research have to be examined. According to Razzaque and Cheng (1998, p 93), Muller (1993) seemed to be the first author to identify different types of 3PLs. To him, there are four distinct classes of 3PLs: (1) asset-based 3PLs use their own assets (e.g. trucks and warehouses) to offer dedicated logistics services; (2) management-based 3PLs supply logistics management through information systems or consulting; (3) integrated 3PLs use their own assets to provide logistics services, but also subcontract from other vendors if necessary; and (4) administration-based 3PLs mainly offer administrative management (e.g. freight payment).

This taxonomy is very similar to that of Africk and Calkins (1994), who distinguish asset-based and non-asset based 3PLs. Other definitions take a different approach by not referring to the resource base of the 3PL, but to the services offered. Berglund et al. (1999, p. 59) suggest that a 3PL is that LSP, which offers "[…] activities […] consisting of at least management and execution of transportation and warehousing […]. In addition, other activities can be included […]. Also, [they] require the contract to contain some management, analytical or design activities, and the length of the co-operation between shipper and provider to be at least one year […]." A similar definition is suggested by van Laarhoven, Berglund, and Peters (2000, p. 426). In line with the definition of third party logistics provided

[10] These terms can be used synonymously, see Knemeyer and Murphy 2004, p. 35.

previously, the present study will not distinguish between asset-based and asset-free providers, but will instead define a 3PL as an LSP that provides strategic, long-term oriented, and not just tactical, short-term oriented services (Knemeyer and Murphy 2005, pp. 5-6) and in the following, the terms 3PL and LSP will be used synonymously.

2.3 Research model

Building on the knowledge on customer loyalty and logistics gained in the preceding sections, this chapter will first detail research needs, before research questions for guiding this dissertation will be derived in the subsequent section. The chapter will close with a presentation of the research approach selected for answering these research questions.

2.3.1 Research needs

As shown in section 2.1.4, a lot of research has been conducted in the field of customer loyalty determinants (see Table 2-1). Scrutinizing those works in detail, however, reveals several deficits:

- Customer loyalty is mostly measured uni-dimensionally, neglecting the differences between repeat purchase, additional purchase, and referrals. Consequently, effects of the studied determinants cannot be separated as to what customer loyalty dimension they influence.
- Very few studies examine more than five determinants of customer loyalty. The approach of limiting analyses to a very small number of determinants is useful in some cases, e.g. when complexity has to be reduced in order to analyze the effects of the selected determinants in detail. When inter-dependencies between these determinants exist, however, this procedure may produce misleading insights, because identified relationships may result only from indirect effects or they can be attributed to factors outside the model. A broader approach therefore appears necessary.
- Finally, problems occur when researchers generalize results although they effectively studied a specific buyer-seller context. In the majority of works, customer loyalty is scrutinized in the consumer goods or industrial equipment sectors. It is highly questionable, whether the derived conclusions are transferable to a buyer-seller relationship in the industrial services context, such as that between LSPs and their customers.

In logistics research, two major deficits can be observed:

- Stemming from the business world, logistics has developed towards an independent research discipline. While research first focused on functional issues, dealing with the question of how to carry out logistics, the management perspective now plays a strong role in logistics research. In addition, the ongoing supply chain management discussion motivated many researchers to examine relationships between different companies. In this field, however, much attention has been paid to third party logistics in general, while "[...] relatively little emphasis has been given to the provider perspective." (Lieb and Randall 1999, p. 28)
- Logistics research oftentimes tries to solve concrete problems from the business world employing either conceptual or qualitative empirical approaches. Even if quantitative empirical studies are undertaken, Mentzer and Kahn (1995) as well as Garver and Mentzer (1999) criticize the lack of theoretical rigor in those works.

Combining the two research disciplines and looking at customer loyalty research in the logistics field, it is evident that very few researchers so far have ventured out to this issue. A comprehensive approach is only provided by Wallenburg (2004), who analyzed customer loyalty for logistics service providers by tackling the following four research questions:

(1) What are the main determinants of customer loyalty regarding repurchase, additional purchase, and referrals?
(2) How can these determinants be integrated into a comprehensive model and what interdependencies can be observed?
(3) Are the effects on repurchase, additional purchase, and referrals sufficiently different to justify the multi-dimensional analysis of customer loyalty?
(4) What is the role of customer satisfaction in creating customer loyalty?

In order to answer these questions, Wallenburg first conceptualized customer loyalty determinants employing a theory pluralistic approach[11]. In this way, he identified 14 factors that potentially influence customer loyalty (Wallenburg 2004, pp. 89-121). Then, he conducted an empirical analysis of German industrial companies using a questionnaire and received 229 usable responses (ca. 4% response rate). The data was then analyzed with structural equation modeling to obtain a causal model of

[11] He based his conceptualization on the following theories: transaction cost theory, exit-voice-theory, social exchange theory, equity theory, risk theory, learning theory, dissonance theory, and commitment-trust-theory (see Wallenburg 2004, pp. 63-83).

customer loyalty and its determinants. In this course, Wallenburg had to eliminate seven factors from his model and ended up with a model consisting of seven factors (proactive improvement, service quality, fairness, price satisfaction, relational satisfaction, alternatives, and commitment), explaining 82.8% of repurchases, 31.2% of additional purchases, and 56.0% of referrals. He also found that the three dimensions of customer loyalty are in fact discriminant and therefore justify detailed analysis (Wallenburg 2004, pp. 205-245). While Wallenburg's work is groundbreaking in customer loyalty research in the logistics field, there are several questions that remain unanswered.

As a consequence of the aforementioned globalization trend, "users are buying logistical services on a more international scale" (Persson and Virum 2001, p. 53) and already in 1993, Cooper identified the importance of intercultural management in logistics (Cooper 1993). To Stone (2001, p. 99), there is no question that "[t]raditional national approaches are giving way to cross border organisations requiring corresponding LSP support" and it is clear that culture influences the way in which people from different countries interact (Adler, Doktor, and Redding 1986, p. 296). Nevertheless, very little research that compares logistics practices in different countries has been conducted yet. The following list gives some examples of comparative research in the logistics field:[12]

- Lieb, Millen, and van Wassenhove (1993) compare logistics outsourcing in the USA and Europe;
- Fawcett and Clinton (1997) compare logistics management practices in Germany, the USA, and Japan;
- Morash and Clinton (1997) compare supply chain organizational structures in the USA, Japan, Korea, and Australia;
- Carter, Ellram, and Ready (1998) compare purchasing behavior in Germany and the USA;
- Sohal et al. (1999) compare quality management practices in the logistics function in Europe, North America and Australia;
- Bookbinder and Tan (2003) compare Asian and European logistics systems and group countries into tiers according to the level of logistics excellence displayed;
- Murphy and Poist (2003) examine environmental logistics in US and non-US firms.

All these studies take a customer's perspective on the challenge of international logistics management. While Grant (2004, p. 183) criticizes that

[12] A limited overview can also be found in Babbar and Prasad 1998.

"specific literature is lacking regarding the differences between national or cultural management decision-making styles in a logistics context", it is even more surprising that no comparative international studies taking the LSP perspective exist yet. Not only customers are confronted with globalizing business, but also LSPs, as "many international buyers increasingly want to consolidate the number of logistics suppliers that they use" (Persson and Virum 2001, p. 54).

Samiee and Walters (2003, p. 194) state that "[u]nlike relationships in domestic settings, those formed across national boundaries are affected to a much higher degree by diverse social, cultural, and other environmental factors which, in turn, can significantly modify the role and scope of relationships from one country or region to the next." Still, their literature review only yields two studies in the field of relationship marketing that involve surveys in two or more countries. And even in those works, comparisons are not made (Samiee and Walters 2003, p. 198). One of the few exceptions is the study by De Wulf, Odekerken-Schröder, and Iacobucci (2001), in which data on consumer relationships is gathered in the USA, The Netherlands, and Belgium. However, only rudimentary comparisons between the three countries are performed. In total, there is hardly any evidence that researchers so far followed the advice of Steenkamp and Baumgartner (1998a, p. 78), who state that the "validity of models developed in one country [should] be examined in other countries".

So far, little attention has been afforded to the contingencies that may have an influence on the formation of customer loyalty in third party logistics relationships. Taking a relationship marketing view on the topic, however, leads to the strong supposition that especially relationship characteristics will have a decisive impact on the way customers are bound to their service providers (Homburg, Giering, and Menon 2003). While a lack of research can be attested for general customer loyalty literature, this deficit is even more pronounced in the logistics context, where the majority of articles is still based on a "one-best-way" paradigm (Chow, Heaver, and Henriksson 1994, p. 26). And even though researchers advise to empirically examine contingency factors in logistics research (Pfohl and Zöllner 1997, pp. 306-318), very few articles have ventured out into this field. A notable exception to this is provided in an article by Murphy, Daley, and Knemeyer (1999), who scrutinize the influence of some general contingencies on logistics management. The only work viewing relationship characteristics as contingencies in the logistics context stems from Knemeyer and Murphy (2005), who analyze the effects of different relational constructs on customer loyalty. However, this study only uses regression analysis to determine direct relationships between relationship characteristics and customer loyalty, and does therefore not offer insights on how relational contingencies moderate a comprehensive customer loyalty model.

2.3.2 Research questions

The preceding section identified several research deficits in the areas of logistics and customer loyalty. In order to tackle these deficits, this section will be devoted to the derivation of a conclusive set of research questions, which will support accomplishing the superordinate goal of this dissertation, i.e. to shed more light on the mechanisms underlying the relationships between LSPs and their customers. Building on the existing, comprehensive research of Wallenburg (2004), two topics were chosen that promise exceptionally interesting insights. The first is concerned with the influence of national culture on the formation of customer loyalty. While it would evidently be desirable to conduct empirical research in a multitude of different countries, such that detailed conclusions could be drawn on the influence of cultural differences, this is impossible to achieve for practical reasons. Instead, two of the largest markets for contract logistics, i.e. Germany and the USA, shall be analyzed. The second research deficit to be tackled here is concerned with the influence of contingencies on the constitution of customer loyalty. While it would again be desirable to examine the moderating effects of a great variety of contingency factors, this study will confine itself to research on moderating effects of relationship characteristics. Within a study nested in the area of relationship marketing, these can be expected to offer specific value for closing some of the stated research gaps.

As a first step of this research, however, the customer loyalty model as proposed by Wallenburg (2004) has to be revisited. Being able to build on Wallenburg's results and experiences, the model shall first be enhanced in certain areas, before it will be examined empirically in Germany and the USA. In this course, the most important determinants will be adopted and the operationalization of constructs will be revised to reflect Wallenburg's findings. Therefore, the first two research questions (RQ) to be answered in this study correspond to Wallenburg's first two questions:

RQ_1: What are the relevant determinants of customer loyalty regarding repurchases, additional purchases, and referrals?

RQ_2: How can the identified factors be integrated into a comprehensive model of customer loyalty and what interdependencies are observed?

Once the developed model is validated empirically in the USA and Germany, moderating effects of relationship characteristics can be analyzed. Knowing exactly how the model of customer loyalty differs as a consequence of contingency factors is of high importance to both theory and practice. This is reflected in the third research question:

RQ_3: *Which relationship characteristics are expected to moderate the formation of customer loyalty between LSPs and their customers and which moderating effects on the model of customer loyalty can be observed?*

Finally, differences between Germany and the USA can be analyzed. Concerning this matter, it is of interest whether differences between the USA and Germany can be hypothesized based on theoretical considerations and on the basis of previous empirical research. These hypotheses shall then be empirically examined. The fourth research question therefore is:

RQ_4: *What differences between the USA and Germany regarding the model of customer loyalty are expected and is there empirical evidence suggesting that these differences actually exist?*

2.3.3 Research approach

In part, the research questions outlined in the preceding section can be answered through conceptual considerations. In order to ensure the validity of the derived hypotheses, however, an empirical examination has to follow. Since a multitude of approaches is feasible for this type of research, the approach used in this study is presented below.

The framework of logistics research proposed by Mentzer and Kahn (1995, pp. 233-240) serves as a basis to the approach employed in this research and is illustrated in Figure 2-5. Idea generation took place prior to starting the research project. Then, integrative literature review was conducted to provide a "historical perspective of the respective research area[s]" and to "formulate a research agenda" (Mentzer and Kahn 1995, p. 233). As a consequence of the literature review, substantive justification is attested for the research project, resulting in the formulation of research questions. Both literature review and substantive justification are given in chapter 2 of this study. Next, the theory-base of the research project is laid in chapter 3. On this foundation, constructs in the form of determinants of customer loyalty and contingency factors are derived in chapter 4. Also, hypotheses are formulated. Then, the applicable methodology is presented in chapter 5, before constructs (measures) are operationalized in chapter 6. Chapter 7 is then devoted to the analysis of the gathered data, where the general model of customer loyalty, moderation analyses, and the effect of national culture are discussed. The dissertation concludes by providing a summary, critically discussing results and finally indicating potentials for future research in chapter 8.

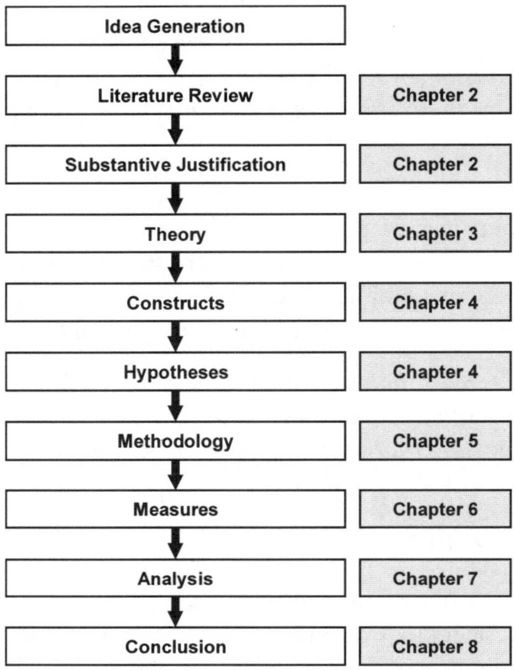

Fig. 2-5. Framework of logistics research[13]

[13] Adapted from Mentzer and Kahn 1995, p. 234.

3 Theoretical framework

In the course of this chapter, the necessary theoretical fundament for this study will be laid. For this, chapters 3.1 to 3.3 will introduce the theories used for conceptualizing a comprehensive explanatory model of customer loyalty (RQ1 and RQ2). Social exchange theory, commitment trust theory, and equity theory will be serving as the basis for deriving potential customer loyalty determinants. While factors stemming from other theories may be suitable for providing additional insights, Wallenburg (2004) came to the conclusion that socio-economic factors have the greatest power for explaining customer loyalty. In fact, an important finding from his study is that customer loyalty does not result from an obligation to remain with a certain LSP, but rather from the desire to do so. Taking into account also the literature review on customer loyalty determinants research conducted in chapter 2.1.4, the present study will exclusively focus on relational factors originating from social exchange theory, commitment trust theory, and equity theory. As each of these theories were already part of Wallenburg's work, only their most important aspects, as well as their explanatory value for conceptualizing customer loyalty determinants, will be briefly discussed.

Section 3.4 will round off the chapter by providing an overview of different cultural concepts and a thorough examination of three influential empirical studies in this field. From this analysis, distinct differences between the USA and Germany will be derived, which will serve as the basis for the formulation of hypotheses on cultural influences on the constitution of customer loyalty, as posited by the fourth research question (RQ4).

3.1 Social exchange theory

3.1.1 Basic concept

Social exchange theory deals with "the relational interdependence, or relational contract, that develops over time through the interactions of the exchange partners." (Lambe, Wittmann, and Spekman 2001, p. 3) While this concept has only entered marketing theory in the 1980s, the discussion goes as far back as to the Greek philosopher Aristotle, who distinguished social exchange from economic exchange in his Nicomachean Ethics (Aristotle 1999, 1162a34-1163a24). Later, sociologists such as Blau (1960, 1964, 1968), Homans (1958), and Emerson (1962), as well as social psychologists such as Thibaut and Kelley (1959) worked on the concept of social exchange.

On the basis of a thorough investigation of social exchange literature, Lambe, Wittmann, and Spekman (2001, p. 6) summarize the following four foundational premises of social exchange theory, which will be detailed below.

- "[E]xchange interactions result in economic and/or social outcomes, […]
- these outcomes are compared over time to other exchange alternatives to determine dependence on the exchange relationship, […]
- positive outcomes over times increase firms' trust of their trading partner(s) and their commitment to the exchange relationship, and […]
- positive exchange interactions over time produce relational exchange norms that govern the exchange relationship."

In contrast to pure economic theories, social exchange theory thus incorporates both economic and social outcomes, highlighting the fact that social as well as economic considerations are made when evaluating the value of relationships. This value results from an assessment of both utilities and costs of a relationship, and parties will choose to uphold a relationship as long as the cost-utility-ratio is satisfactory. The exact composition of utilities and costs varies from individual to individual, and more emphasis may be put on either social or economic aspects. Nevertheless, utilities and costs are not evaluated separately, but jointly, as pointed out by Homans (1958), Thibaut and Kelley (1959), and Blau (1964).

This cost-utility-ratio, which is the actual benefit (B) realized in a relationship, is then compared to some standard to assess the beneficialness of a relationship. A conceptualization of this comparison is provided by Thibaut and Kelley (1959), who establish a comparison level (CL) and a comparison level of alternatives (CL_{alt}). CL corresponds to the maximum potential benefit that an individual deems feasible to obtain from a relationship, i.e. the benefit that an individual feels entitled to. The more the actual benefit (B), i.e. the cost-utility-ratio realized in the relationship, exceeds CL, the higher is the satisfaction with the relationship. Relationship satisfaction, however, is not sufficient for deciding on whether to continue or expand a relationship or not. For this reason, B is also compared to CL_{alt}, i.e. "the overall benefit […] available from the best possible alternative exchange relationship." (Lambe, Wittmann, and Spekman 2001, p. 9) Only if CL_{alt} is surpassed by the current relationship, will the parties remain in the relationship. Accordingly, a relationship may be continued despite of relationship dissatisfaction, if there are no superior alternatives.

When appraising the value of a relationship, parties will not only consider current and past costs and utilities, but also potential future benefits, and may forego present benefits for benefits in the future. Therefore, an important factor in social exchange theory is trust, which results from multiple and beneficial interactions over time (Blau 1964). Only if the trading partner is trusted, will one be willing to reciprocate the abdication of current benefits for future benefits (Homans 1958). Another important aspect of social exchange theory is commitment, which is fostered by trust. This causal relationship between trust and commitment is based on the principle of generalized reciprocity, because "commitment entails vulnerability [and] parties will seek only trustworthy partners". (Morgan and Hunt 1994, p. 24)

Social exchange theory's foundational premises indicate that relationships are not only governed by contracts, but also by norms, which develop as a consequence of repeated interaction (Thibaut and Kelley 1959) and "increase the efficiency of relationships because by agreeing to the manner in which interactions take place, the degree of uncertainty may be reduced." (Lambe, Wittmann, and Spekman 2001, p. 12) As with trust, norms are adhered to because rewards are expected.

While social exchange theory can be very useful conceptually, it is mainly criticized for its lack of empirical foundation. Factors such as commitment and trust should be able to substitute (at least in part) contractual governance structures (Griesinger 1990, p. 488), but this cannot be confirmed empirically (Rindfleisch and Heide 1997, p. 50). Presumably, this is attributable to social exchange theory's neglect of opportunism. Moorman, Zaltman, and Deshpandé (1992, p. 323) note that dependence and vulnerability inherent in close relationships may eventually lead to the perception that one party takes advantage of the other, creating the need for formal governance structures that can serve as safety nets.

3.1.2 Explanatory power regarding customer loyalty

As with other settings, relationships between LSPs and their customers are not only characterized by economic, but also by social considerations. For this reason, social exchange theory can be applied to shed light on these factors that are typically neglected by classic economic theories. It is clear that customer loyalty in particular is created by a combination of economic and social aspects.

There already is a substantial body of literature that proposes relational success factors on the basis of social exchange theory. A comprehensive overview of works incorporating these factors is given by Lambe, Wittmann, and Spekman (2001, p. 16-19). Revisiting the CL/CL_{alt} paradigm proposed by Thibaut and Kelley (1959), it becomes clear that the implied decision is that of loyalty. High satisfaction, resulting from the comparison between CL and the actual benefit achieved in the relationship, makes relationship continuation more likely, as satisfied parties are less likely to leave a relationship (Anderson and Narus 1984, p. 70). Since the final loyalty decision is made by comparing B with CL_{alt}, market alternatives play an important role and will be used as a comparison level. In addition, trust and commitment increase the perception of achieved benefits, because present benefits may be foregone in hope of future benefits and therefore augment the likelihood of relationship continuation. In addition, service providers can demonstrate their dedication to the relationship by continuously and independently pursuing service improvements. Both the implied signaling effect and the actual performance melioration can nurture the buyer's perception of gained benefit and future prospects. Social exchange theory therefore provides the following five factors that are expected to influence customer loyalty: satisfaction, alternatives, trust, commitment, and proactive improvement.

3.2 Equity theory

3.2.1 Basic concepts

Equity theory represents an extension of social exchange theory by adding the aspect of fairness[1]. While the concept dates back to Homans (1958), equity theory was primarily coined by Adams (1963, 1965). At that time referred to as "theory of inequity", it was introduced to explain wage inequities. The basic assumption underlying equity theory is that each party in a relationship compares its input-output-ratio to that of the other party. Analogous to social exchange theory, social as well as economic considerations are incorporated in the evaluation of fairness. In case the ratio is balanced, the perception of being fairly treated is conveyed. Otherwise, one feels unfairly treated, arousing distress for both the over-benefited and

[1] Instead of the word "equity", "fairness" is used in this study, as "fairness" is the term usually referred to by LSPs and their customers.

the under-benefited parties (Austin and Walster 1975, p. 475), which may lead to emotional and behavioral consequences. In order to re-create fairness, Wallenburg (2004, p. 80) proposes that parties can either change their inputs, adjust their expectations, influence the other party, or terminate the relationship.

As in social exchange theory, the input-output-ratios of the involved parties do not have to be in balance at any given time. Instead, parties must trust that outcomes be split equitably in the long-run (Lambe, Wittmann, and Spekman 2001, p. 9).

3.2.2 Explanatory power regarding customer loyalty

Equity theory postulates a fair split between inputs and outputs across firm boundaries, acknowledging that inequitable distributions may lead to relationship termination (Huppertz, Arenson, and Evans 1978, p. 250). As such, equity theory is one of the few economic theories, which explicitly incorporates inter-firm relationships, abstaining from focusing on a single firm only. Equity theory is therefore useful for conceptualizing fairness as a determinant of customer loyalty.

In addition, logistics and supply chain researchers agree that firms can only cooperate successfully if both risks and benefits of the relationship are fairly split (e.g. Wallenburg 2004, p. 81). The inclusion of equity theory and the concept of fairness into this customer loyalty study in the logistics context is therefore appropriate and promises to provide additional explanatory value for the intended comprehensive customer loyalty model.

3.3 Commitment trust theory

3.3.1 Basic concepts

Morgan and Hunt (1994) first introduced commitment trust theory in their article on successful relationship marketing. According to them, commitment and trust function as key mediating variables between five antecedents (relationship termination costs, relationship benefits, shared values, and opportunistic behavior) and five outcomes (acquiescence, propensity to leave, cooperation, functional conflict, and decision-making uncertainty). By highlighting commitment and trust, Morgan and Hunt's theory is based on the fundamental ideas of social exchange theory. At the same time, one of the major deficits of social exchange theory is addressed by allowing for opportunistic behavior.

Considerations by Morgan and Hunt were spurred by a shift in marketing research and practice away from a mere transactional focus towards the relationship marketing approach, according to which all marketing activities are supposed to establish, develop, and maintain successful relational exchanges (Morgan and Hunt 1994, p. 22). While Morgan and Hunt acknowledge that many contextual factors determine the success or failure of relationship marketing efforts, commitment and trust are seen as key, because they can establish relational governance norms. As such, commitment and trust can encourage cooperative behavior aimed at preserving relationship investments, mitigate the risk of choosing attractive short-term alternatives despite of expected long-term benefits with existing exchange partners, and can make high-risk actions appear more attractive, because exchange partners are not feared to act opportunistically. "Therefore, when both commitment and trust – not just one or the other – are present, they produce outcomes that promote efficiency, productivity, and effectiveness." (Morgan and Hunt 1994, p. 22)

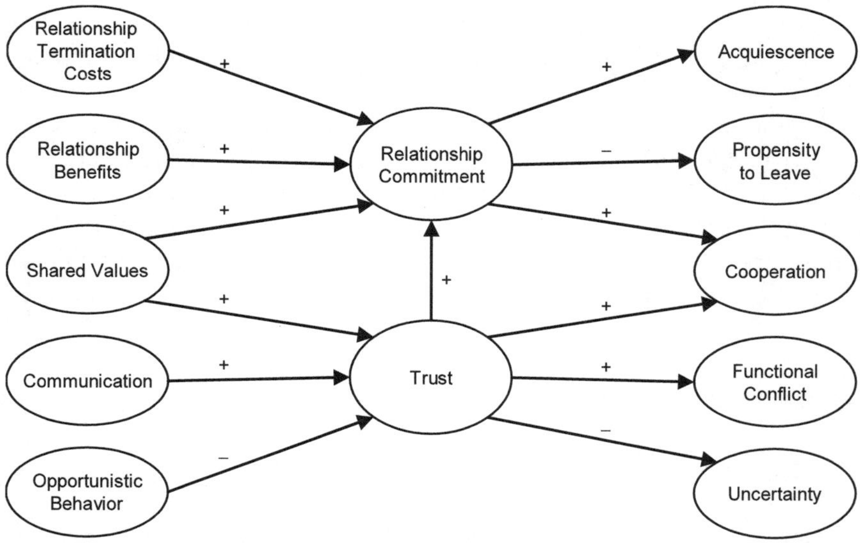

Fig. 3-1. Morgan and Hunt's key mediating variables model[2]

The above mentioned antecedents and outcomes are integrated into a causal model (see Figure 3-1), which also includes a positive relationship between trust and commitment. This is in line with social exchange theory, in which trust over time is supposed to nurture commitment. This model

[2] Morgan and Hunt 1994, p. 22.

was tested empirically on a sample of 204 automobile tire retailers and results supported Morgan and Hunt's hypotheses.

3.3.2 Explanatory power regarding customer loyalty

Researchers of marketing and relationships generally agree on the importance of commitment and trust for the success of relationships (Morris, Barnes, and Lynch 1999, p. S660). Great value is created by Morgan and Hunt through the formulation of a comprehensive model and its empirical validation and as a result, their article is one of the most cited in recent relationship research.

With respect to customer loyalty, Morgan and Hunt's commitment trust theory is relevant by proposing two factors that foster successful relationships. In addition, repurchase is explicitly included in their model (propensity to leave) and shown to be directly influenced by commitment and only indirectly by trust.

3.4 Cultural theory

This chapter is dedicated to the identification of cultural differences between Germany and the USA. First, the concept of culture will be introduced to create a notional base for the following deliberations. Then, three complementary models of culture will be introduced. The section will close by giving a summarizing overview of differences between Germany and the USA that can be expected on the grounds of cultural theory.

3.4.1 Introduction to the concept of culture

Culture has been subject to ongoing research during the past centuries, especially in disciplines such as philosophy, anthropology, biology, psychology, and economics. However, a multitude of definitions exists for the notion of culture and a common understanding is yet to be established.

One of the most frequently cited definitions of culture is provided by Kluckhohn (1951, p. 86), who states that "[c]ulture consists in patterned ways of thinking, feeling and reacting, acquired and transmitted mainly by symbols, constituting the distinctive achievements of human groups, including their embodiment of artifacts; the essential core of culture consists of traditional (i.e. historically derived and selected) ideas and especially their attached values." Clearly, the main deficits of this definition are its broadness and lacking specificity, which impede its direct application to the concrete questions at hand in this work. Another frequently cited definition is given by Hofstede (2001), which is based on the concept of 'habitus' proposed by Bourdieu (1980)[3]. According to Hofstede, culture influences an individual's "mental programming" (Hofstede 2001, p. 2), or 'habitus', of which three layers exist: the universal level is the least unique, because it is common in all human beings and constitutes our "operating system" (p. 2). In this sense, the universal level accounts for general traits, characteristics, and behaviors "such as laughing and weeping, associative and aggressive behaviors that are also found in higher animals." (p. 2) The second level, termed collective level, "is shared with some but not all other people; it is common to people belonging to a certain group or category, but different from people belonging to other groups or categories." (p. 2) The third level is called the individual level, because it is unique in every human being, pertaining to an individual's own personality. While the first and third levels are either common in all human beings or different between each and every person, the collective level is most likely to be shared by individuals, "who have gone through the same learning processes" (p. 3). Consequently, Hofstede (2001, p. 9) defines culture as "the collective programming of the mind that distinguishes the members of one group or category of people from another." Figure 3-2 provides an overview of Hofstede's human mental programming model.

While this definition by Hofstede serves as the conceptual basis of the notion of culture assumed in this work, it is apparent that culture cannot be easily assessed on this ground alone. Additionally, this definition is not sufficiently specific for national culture, wherefore two further models, one by Hofstede (2001, p. 11) himself, and another one posited by Trompenaars and Hampden-Turner (1997, p. 22), will be presented to concretize the notion of national culture. Both Hofstede and Trompenaars and Hampden-Turner formulate 'onion-models' consisting of several distinct layers, differing in the degree of direct observability (see Figure 3-3). In both models, the outer layer is directly observable, denoted "symbols" by the former author, and "artifacts and products" by the latter authors. The central layer for both models is not observable, as it comprises "values" (Hofstede 2001) or "basic assumptions" (Trompenaars and Hampden-Turner 1997) and remains "invisible until [it] becomes evident in behav-

[3] Hofstede 2001 translates Bourdieu's remarks on 'habitus' in the following way: "Certain conditions of existence produce a habitus, a system of permanent and transferable tendencies. A habitus...functions as the basis for practices and images...which can be collectively orchestrated without an actual conductor" (p. 3).

ior" (Hofstede 2001, p. 10). In between these layers, Hofstede proposes two further layers, "rituals" and "heroes". While "rituals" are e.g. expressed through "[w]ays of greeting or paying respect to others" (p. 10), "[h]eroes are persons, alive or dead, real or imaginary, who possess characteristics that are highly prized in a culture and thus serve as models for behavior." (p. 10) Notwithstanding from this, Trompenaars and Hampden-Turner (1997, p. 22) only propose one intermediate layer, called "norms and values", which encompasses desirable behavior as deemed by a culture. Both models are shown in Figure 3-3.

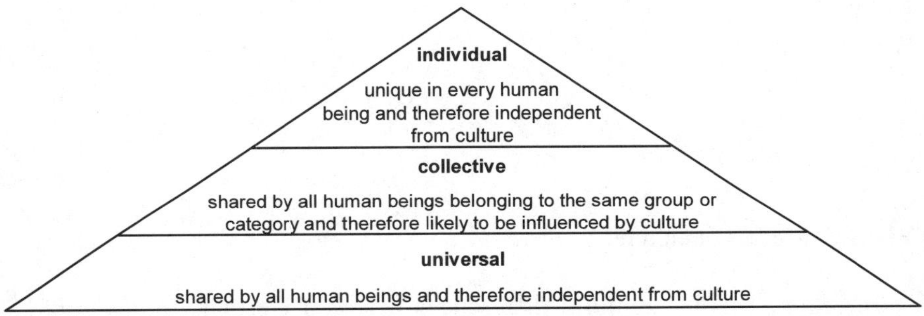

Fig. 3-2. Human mental programming[4]

At the core of these concepts lies the assumption that implicit values form cultures' members' attitudes, leading to specific behavioral patterns (compare also Kluckhohn 1951 and Hall and Hall 1990). Hence, if cultural values inherent in a certain culture are known, human behavior can, at least to some degree, be predicted.

Additionally, national culture and other types of culture, e.g. corporate culture (i.e. the culture within a specific company) or professional culture (i.e. the culture of members of a specific profession or job) have to be distinguished (Shane 1995, p. 52, Luk 1997, pp. 13-14, Trompenaars and Hampden-Turner 1997, p. 7, Hofstede 2001, p. 10). While it may be argued that national culture is loosing significance in an increasingly globalizing world, researchers are still able to empirically show significant differences in national cultures (e.g. Adler and Graham 1989). Consequently, in the context of this research, national culture will be understood as the collective programming of the mind that distinguishes the members of one national culture from another (adapted from Hofstede 2001, p. 9), i.e. "that component of our mental programming which we share with more of our compatriots as opposed to most other world citizens." (Hofstede 1989, p. 193)

[4] Adapted from Hofstede 2001, p. 3.

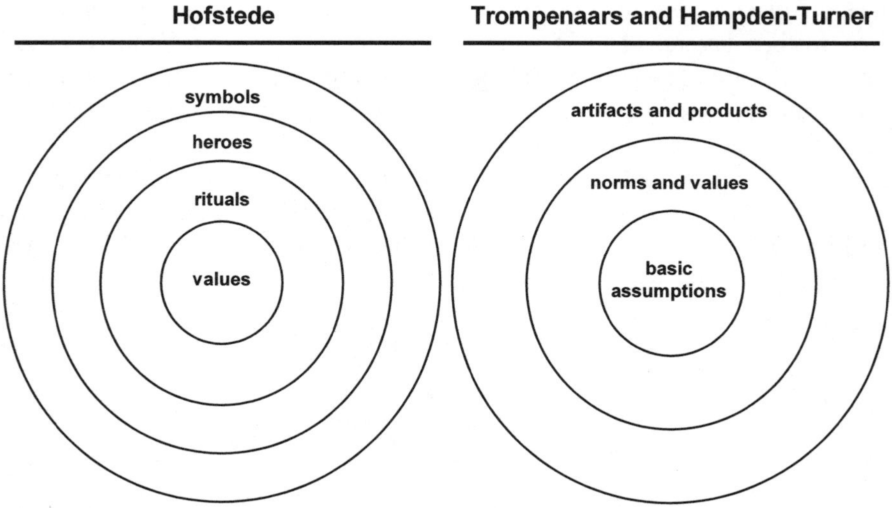

Fig. 3-3. Models of culture[5]

In order to assess cultural differences between Germany and the USA, the concept of culture has to be operationalized. According to Singh (2004, p. 95), culture may be assessed on the basis of cultural values (e.g. Kluckhohn and Strodtbeck 1960, Hall 1976, Hall and Hall 1990, Hofstede 1980/2001, Trompenaars and Hampden-Turner 1997), or on the basis of cultural norms, routines, or customs (e.g. Goodenough 1981, D'Andrade 1984, McCort and Malhotra 1993). Since none of the mentioned approaches appear to present a complete view on national culture, the three most-cited studies will be introduced in the following sections. These frameworks are applicable in the context of the present research, as they are complimentary and include explicit consideration of both Germany and the USA.

3.4.2 Hofstede's framework of cultural differences

With more than 100,000 valid questionnaires from 53 countries, the empirical study conducted by Hofstede (1980) remains the largest of its kind. After joining IBM in 1965 as a human resources researcher, Hofstede conducted IBM's international employee attitude survey between 1967 and 1973. Since IBM operated offices in approximately 100 countries at that time, Hofstede had the opportunity to obtain data from 66 IBM offices worldwide. In his original study published in 1980, Hofstede measured na-

[5] Adapted from Hofstede 2001, p. 11 and Trompenaars 1996, p. 51.

tional culture along four distinct dimensions (power distance, uncertainty avoidance, individualism/collectivism, masculinity/feminity) that will be outlined below. As he became aware of the fact that this study included a western-hemisphere bias, another study was conducted in 23 Asian and non-Asian countries to include Asian cultural aspects. This data was combined with the original data and analyses yielded a fifth cultural dimension, long-term vs. short-term orientation (Hofstede 2001)[6]. These dimensions and the empirically validated differences between Germany and the USA along these dimensions will be outlined in section 3.4.2.1. Particular aspects distinguishing Germany and the USA will be detailed in sections 3.4.2.2 and 3.4.2.3 and a summary and critical assessment will be provided in section 3.4.2.4.

3.4.2.1 Dimensions of culture and cultural distance

Power distance (PD), as defined by Hofstede (2001, pp. 79-143), refers to the treatment of human inequality in a society and is shown to be considerably influenced by a society's collective programming. "Inequality can occur in areas such as prestige, wealth, and power" (p. 79) and "is usually formalized in boss-subordinate relationships." (p. 79) In this context, Hofstede (2001, p. 83) provides the following rather tangible definition: "Power distance between a boss B and a subordinate S in a hierarchy is the difference between the extent to which B can determine the behavior of S and the extent to which S can determine the behavior of B." Hence, PD refers to the distribution of power in the relationship between boss and subordinate, where PD is the higher the more power the boss has over his subordinate, in comparison to the power the subordinate has over its boss.

Quantitatively, Hofstede bases his assessment of PD on answers to three questions included in his original study with IBM data. Those questions'[7] mean scores per country were quantitatively related to form a power distance index (PDI), which can be stated formally as:

"PDI = 135 – 25(mean score employees afraid) + (percentage perceived manager 1+2) – (percentage preferred manager 3, 1967-1969)"
(Hofstede 2001, p. 86),

where the number '135' was introduced to normalize PDI to a range between -90 and +210, where usual values fall between zero (low PD) and 100 (high PD). In his empirical study, Hofstede (2001, p. 87) ranks 53

[6] See Hofstede 2001, pp. 41-77 for a description of data collection procedures and methodology used.
[7] See Hofstede 2001, pp. 470, 472 for the actual questions.

countries along PDI, with values ranging from 11 (lowest PD, Austria) to 104 (highest PD, Malaysia), an average score of 57, and a standard deviation of 22. Germany and the USA score at values of 35 and 40 respectively, indicating that both countries clearly are countries demonstrating low PD. The results are graphically shown in Figure 3-4 on page 51. Consequently, cultural differences on the basis of PD cannot be expected and a detailed analysis of the PD dimension is not necessary.

The second dimension, uncertainty avoidance (UA), as defined by Hofstede (2001, pp. 145-208), captures the extent to which members of a society feel threatened by uncertainty regarding the future. It is assumed that different cultures react in different ways to reduce the threats posed by uncertainty, e.g. through law, i.e. a society's formal and informal rules, or through religion. In short, Hofstede (2001, p. 161) defines UA as "[t]he extent to which the members of a culture feel threatened by uncertain or unknown situations."

In order to measure UA, Hofstede (2001, pp. 149-150) introduces the uncertainty avoidance index (UAI), which is based on answers to three questions[8] dealing with rule orientation, employment stability, and stress. In formal representation, UAI can be stated as:

"UAI = 300 – 30(mean score rule orientation) – (percentage intending to stay less than 5 years) – 40(mean stress score)" (Hofstede 2001, p. 150),

where the constant '300' was introduced to normalize UAI. Empirically, Hofstede (2001, p. 151) determined UAI scores for 53 countries, ranging between 8 (lowest UAI, Singapore) and 112 (highest UAI, Greece), with a mean of 65 and a standard deviation of 24. While the USA show a low degree of UA (score of 46), Germany is positioned right at the mean with a score of 65. Consequently, Germany cannot be labeled a low-UAI country, but will rather show traits of both low- and high-UAI societies. The USA, on the other hand, are strictly low-UAI and can therefore be expected to differ significantly from Germany. The scores are graphically represented in Figure 3-4 on page 51 and further information on cultural differences between the USA and Germany will be detailed in section 3.4.2.2.

By individualism and collectivism (I/C), Hofstede (2001, pp. 209-278) denotes the degree of gregariousness prominent in a society. "[T]he relationship between the individual and the collectivity that prevails in a given society [...] is reflected in the way people live together" (Hofstede 2001, p. 209). The fundamental aspect of the I/C dimension therefore is the dominant self concept within a society, i.e. whether individual matters are prioritized opposite matters that concern the society as a whole. Simply

[8] See Hofstede 2001, pp. 469, 472 for the actual questions.

put, it is the issue whether individuals primarily look after themselves, preferring loose-knit social networks, or if tightly-knit social networks are preferred, in which members of society look after one another (Hofstede 1985, p. 348).

In order to measure I/C, Hofstede (2001, p. 214) introduces the individualism index (IDV), which is based on 14 general questions[9] on work goals. IDV is then determined using factor analysis, which resulted in two factors: one stressing independence from the organization, and another stressing dependence. An index was calculated by correlating the individual work goals to the country mean scores for the two factors, and normalizing the index between zero (very collectivistic) and 100 (very individualistic). On the basis of his data, Hofstede (2001, p. 215) computed IDV scores for 53 countries, ranging from 6 (lowest IDV score, Guatemala) to 91 (highest IDV score, USA), with a mean of 43 and a standard deviation of 25. While the USA clearly came out as the most individualistic country in the sample, Germany scored a 67, which is well above the mean, but still significantly less than the USA. Hence, the USA can be expected to show individualistic traits exclusively, while Germany will still show some collectivistic traits. Again, the results are graphically shown in Figure 3-4, p. 51 and details of resulting cultural differences will be given in section 3.4.2.3.

The cultural dimension of masculinity and femininity (M/F), outlined by Hofstede (2001, pp. 279-350), is concerned with the duality of the sexes and the specific ways different cultures have developed to cope with it. Hofstede (1985, p. 348) asserts that "masculinity [...] stands for a preference for achievement, heroism, assertiveness, and material success, as opposed to femininity, which stands for a preference for relationships, modesty, caring for the weak, and the quality of life." Hofstede summarizes the male role as 'tough', and the female role as 'tender', and concludes that "[i]n a masculine society even the women prefer assertiveness (at least in men); in a feminine society, even the men prefer modesty." (Hofstede 1985, p. 348) Thus, Hofstede (2001, p. 297) defines: "Masculinity stands for a society in which social gender roles are clearly distinct: Men are supposed to be assertive, tough, and focused on material success; women are supposed to be more modest, tender, and concerned with the quality of life. Femininity stands for a society in which social gender roles overlap: Both men and women are supposed to be modest, tender, and concerned with the quality of life."

[9] See Hofstede 2001, pp. 467-468 for the actual questions.

Using his original data, Hofstede (2001, p. 279) again conducted factor analysis on mean country work goal scores[10] to obtain a masculinity index (MAS). As with IDV, MAS was normalized to values between 0 (purely feminine) and 100 (purely masculine). From his data, Hofstede (2001, p. 286) computed MAS scores for 53 countries, ranging from 5 (lowest MAS, Sweden) to 95 (highest MAS, Japan), with a mean of 49 and a standard deviation of 18. Both Germany (score 66) and the USA (score 62) are clearly masculine societies and significant differences between the two countries on the basis of the M/F dimension cannot be expected, wherefore a detailed analysis of the M/F dimension is not necessary. The scores of the M/F dimension are depicted in Figure 3-4, p. 51.

Hofstede (2001, pp. 351-372) introduced the fifth dimension of long- versus short-term orientation (L/S) after it became apparent that his original survey using IBM data was beclouded by a western bias, attributable to the survey's designers' mainly western descent. On the basis of a study conducted by Asian scholars in 1985 on 2,300 students in 23 countries from all over the world, Hofstede (2001) validated his fifth dimension. L/S is reminiscent of some of Confucius' teachings and refers to a general short- or long-term orientation in life. To the point, Hofstede (2001, p. 359) defines: "Long Term Orientation stands for the fostering of virtues oriented towards future rewards, in particular, perseverance and thrift. Its opposite pole, Short Term Orientation, stands for the fostering of virtues related to the past and present, in particular, respect for tradition, preservation of 'face' and fulfilling social obligations."

Measurement of L/S is based on factor analysis on eight values ('persistence', 'ordering relationships by status and observing this order', 'thrift', and 'having a sense of shame' indicate long-term orientation, while 'personal steadiness', 'protecting your face', 'respect for tradition', and 'reciprocation of greetings, favors, and gifts' form short-term orientation). The factor scores for the surveyed countries were transformed into a long-term orientation index (LTO) that was originally normalized to values between zero and 100, where 100 indicates extreme long-term orientation, and zero extreme short-term orientation (Hofstede 2001, pp. 354-355). LTO scores were computed for 23 countries, ranging from zero (lowest long-term orientation, Pakistan) to 118 (highest long-term orientation, China), with a mean of 46 and a standard deviation of 28. Both Germany (score 31) and the USA (score 29) are clearly below average and can therefore be labeled rather short-term oriented. As the difference between the countries is minute, however, significant differences as a result of L/S are not to be ex-

[10] See Hofstede 2001, pp. 467-468 for the actual questions.

pected and a detailed analysis of the L/S dimension is not conducted. The scores are again graphically represented in Figure 3-4.

It is evident that little differences between Germany and the USA exist on three of the five dimensions (power distance, masculinity/femininity, long- vs. short-term orientation), while significant differences are revealed regarding the remaining two dimensions (uncertainty avoidance, individualism/collectivism). The same results can be obtained algebraically by computing the cultural distance index as proposed by Kogut and Singh (1988) and Slangen and Hennart (2003).

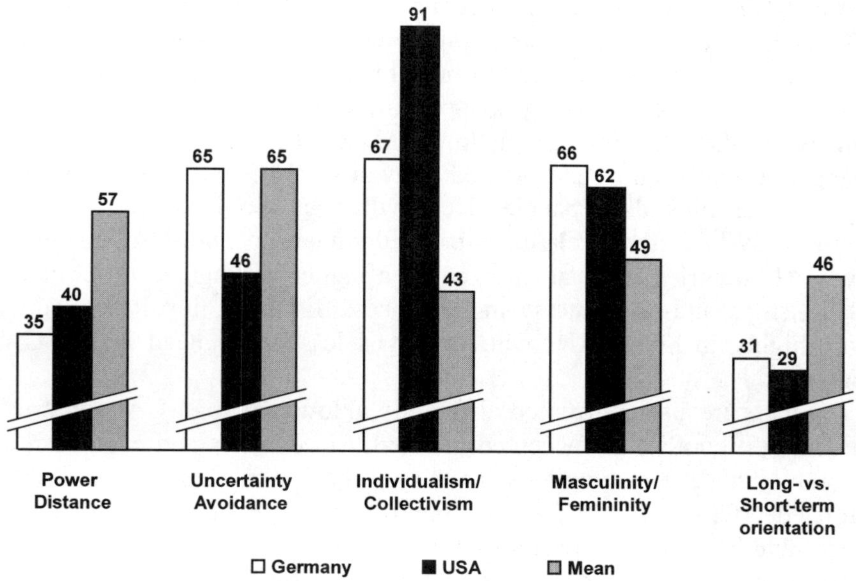

Fig. 3-4. Germany and the USA on Hofstede's five dimensions

3.4.2.2 Uncertainty avoidance

Most importantly, the degree of UA in a society leads to differing ways of handling ambiguities and to diverging needs regarding the predictability of everything encompassing an individual. In the context of this study, the following aspects are of highest significance.

Risk acceptance is a major differentiator between low- and high-UAI countries. While members of all societies certainly accept a certain amount of risk, members of low-UAI societies are willing to accept unknown risks, while members of high-UAI countries only accept known risks (Hofstede 2001, p.160). In this sense, known risks are the ones that can be specifically foreseen, such as the danger of getting injured in a car-race, while unknown risks are less specific, like those that result from immersing into a business relationship with an unknown supplier. In consequence, low- and high-UAI countries also demonstrate different levels of appreciation of change and innovation, as these are seen as direct antecedents to unknown risk. Thus, low-UAI countries exhibit higher *appreciation of change and innovation*, while high-UAI countries are hesitant towards change and innovation.

Another difference between high- and low-UAI countries appears with respect to *laws and rules*. High-UAI countries have a high need for rules and laws and try to formalize procedures for every contingency (Hofstede 2001, p.174). Low-UAI countries on the contrary have a smaller need for laws and rules and value freedom and liberty. Also, high-UAI countries observe rules more strictly and oppose any changes to the legislative system, while the view prevails in low-UAI countries that rules should be changed, if they do not seem adequate any more (Hofstede 1983, p. 61).

In relationships, discrepancies occur with regard to the *stability of relationships*. While stable relationships predominate in high-UAI countries, low-UAI countries are characterized by a higher willingness to terminate relationships and to immerse in new ones. Similarly, loyalty is valued more highly in high-UAI countries and is less pronounced in low-UAI countries.

Conflicts are also confronted differently in low- and high-UAI countries. Dissent and conflict are accepted in low-UAI countries and even seen as favorable in certain situations, as they may foster constructive change (Hofstede 1983, p. 61). High-UAI countries, on the other hand, value consensus and try to avoid any sort of open conflict.

Correlating his results to a study conducted by Ornauer et al. (1976), Hofstede (2001, pp. 175-176) reveals that members of low-UAI societies are more willing to *compromise* with opponents than are members of high-UAI countries. This is accompanied by a higher willingness to *trust* others in low-UAI countries, while trust is seen as a potential danger in high-UAI countries. According to Hofstede (2001, p. 159), trust may expose the trustor to the danger of losing control, a state not desired by members of high-UAI societies.

3.4.2.3 Individualism and collectivism

According to Puck (2002, p. 8), "individualism is manifested in high levels of autonomy of employees, in an employee-employer-relationship that is predominantly based on the employment contract, and in work conditions that provide employees with sufficient personal time. Collectivism (or low individualism), on the other hand, is manifested in work unit solidarity, group responsibility for results, moralistic/family-like relationships with employers, and the priority of relationships over tasks." The most important effects of I/C in the context of this study are delineated below.

A clear *distinction between in- and out-group members* is made by and accepted in low-IDV countries, where in-groups could e.g. be members of a specific department or company. On the other hand, members of high-IDV societies demand equal treatment of everyone (Hofstede 2001, p. 238). As a consequence, high-IDV societies apply the same value standards to everybody, while members of low-IDV societies attribute different value standards to in- and out-group members, such that trust is more willingly extended to in-group members than to out-group members. Also, members of high-IDV countries are more likely to reconsider group membership, and may do so every time they are assigned new tasks. In contrast, low-IDV groups are more settled and members are less likely to change their assessment of group membership. Furthermore, low-IDV societies require the other party to become a member of the in-group before engaging in a business relationship. Hence, a requirement for the formation of business relationships is the a-priori creation of a personal relationship with the other party involved, because only then will the other party be accepted to the in-goup (Hofstede 2001, p. 239).

As a result of this, *loyalty* towards the in-group is high in low-IDV countries and relationships within the in-group are likely to be very close, with feelings of "mutual obligations of protection [towards in-group members] in exchange for loyalty." (Hofstede 2001, p. 237) On the other extreme, high-IDV countries have a transaction-based view of relationships and are therefore more prone to end a relationship if the desired outcome is not achieved.

Accordingly, business relationships in high-IDV countries are rather *impersonal*, while they are almost family-like in low-IDV countries (Hofstede 2001, pp. 238-239). In consequence, "[t]o the collectivist mind, only natural persons are worthy of trust [...], but not impersonal legal entities like companies." (Hofstede 2001, p. 239)

Finally, the attribution and causes of *success* differ between low- and high-IDV countries. While success is usually measured on a group basis in low-IDV countries, success measurement on the individual level is common in high-IDV countries. This leads to a more active self-concept in high-IDV countries, where individuals "are more active in trying to get somewhere." (Hofstede 2001, p. 233) Members of low-IDV societies, on the other hand, prefer involving other group-members in important decisions to be sure of their support. As a result, collectivist societies believe in "sharing information, openly committing oneself, and political alliances"

(Hofstede 2001, p. 244), while the opposite is deemed crucial for the achievement of success by members of individualist societies.

3.4.2.4 Summary and critical assessment

While Germany's score on uncertainty avoidance conforms to the average of the 53 surveyed countries, the USA display a very low score. Hence, the USA do not fear change and unknown risks, embrace innovations and exhibit an open attitude. Germans, on the contrary, generally mistrust unfamiliar behavior, change, and innovation, value loyalty and stable business relationships, foster information sharing, seek consensus, and avoid conflict. An important aspect of German business relationships, however, is that they usually require personal bonding as a basis for enduring business relationships, which finds empirical support in the work of Williams, Han, and Qualls (1998).

On the individualism/collectivism dimension, both Germany and the USA are more individualist than the average, yet the USA are much more extreme compared to Germany. Therefore, the USA stress opportunity taking-behavior, perceive business relationships as impersonal, and subscribe to an active self-concept. Germans, on the other hand, display some collectivist traits and ascribe more importance to personal bonds and to preferential treatment of and responsibility for in-group members. As such, they are loyal and trustful towards in-group members, seek consensus, dread any change to the status-quo, and establish rules to maintain a desired state.

Hofstede's work has been praised by a multitude of researchers, celebrating it as a "revolution within the social sciences" (Eckhardt 2002, p. 89). According to the 'Social Science Citation Index', Hofstede is indeed the most-cited Dutch author and the ninth most-cited European author (Bing 2004, p. 80). In the early 1980s, Hofstede's original book on culture's consequences (Hofstede 1980) was the first truly large-scale exploration of culture – until today, there exists no study in the field of cultural studies that is based on a larger database. In addition, no comprehensive framework tries to detect cultural differences and their effects in so many fields as Hofstede's, which deals with issues of family, schools and educational systems, work and organization, political systems, religion, ideology, and theories of power. Also, the applicability of Hofstede's framework to the context of buyer-seller relationships is confirmed by a number of authors (Pressey and Selassie 2003, p. 356) and the numerous replications of (at least parts of) Hofstede's framework (Hoppe 2004, p. 77) further underline its importance in the field of cultural studies. However, Hofstede's work has also been subject to criticism by many researchers. In fact, Hofstede himself discusses the major points of criticism in his second edition (Hofstede 2001, p. 73).

A very often uttered deficit of Hofstede's framework is its limitation to the IBM corporation and its worldwide subsidiaries. While Hofstede (2001, p. 73) claims that his samples always matched the respective countries' whole populations and are thus representative, some researchers disagree. McSweeney (2002, p. 110), for instance, claims that Hofstede's "surveys were confined to certain categories of IBM employees – thus excluding blue-collar workers, the non-employed, the retired, the unemployed, full-time students, the self-employed, and others". The point of this criticism is two-fold. On the one hand, Hofstede restricted his study to employees of one company – a deficit that has to be kept in mind when using the framework. On the other hand, the presented statement criticizes that Hofstede did not actually explore cultural differences, but differences between workplaces in different countries. While this second point of criticism is certainly valid and must be handled with caution in certain circumstances, it is of no major concern for the purposes of this research, as the study of customer loyalty at hand is also confined to Hofstede's sample characteristics – the business environment.

Another frequently expressed doubt about the validity of Hofstede's findings is concerned with the up-to-dateness of his data, which was gathered some 30 years ago. Although Hofstede (2001, p. 73) counters that the identified cultural dimensions "have centuries-old roots", Eckhardt (2002, p. 93) remarks that anthropologists would not support the view that values are inherently stable. This issue is especially relevant for studies involving Germany, as Hofstede exclusively gathered data from the Federal Republic of Germany (FRG) before its reunification with the German Democratic Republic (GDR) in 1990. Given the differences between the political and social systems of the FRG and the GDR that must have certainly influenced collective programming in the two territories, the absolute validity of Hofstede's data cannot be taken for granted in this study focusing on differences between (reunited) Germany and the USA.

Another point criticized (e.g. Goodstein 1981) is the small sample size in some of the surveyed countries, e.g. Pakistan, casting doubt on the validity of statistical methods used by Hofstede. While this also is a valid point, Hofstede's studies in several countries had more than 1,000 responses each, as for instance in Germany and the USA. The proposed results should therefore hold true at least for these countries.

For each of his five dimensions, Hofstede (2001) tries to validate his findings by correlating his results to other studies. This practice is heavily criticized by Eckhardt (2002, p. 91), who notes that Hofstede neglects the many studies that would have challenged his framework. She thereby accuses Hofstede of painting a very biased picture of his results. In line with this, she also points out "that Hofstede holds onto a static vision of culture, a view no longer tenable in light of decades of social science research suggesting otherwise." (Eckhardt 2002, p. 93) In order to counteract this severe criticism, a second empirical framework of cultural differences will be presented subsequently in this study (see section 3.4.4), in which Trompenaars and Hampden-Turner (1997) present a dynamic view of culture.

Despite the multitude of limitations brought up by critics of Hofstede's research, a study of similar scope and applicability is yet to be developed. However, this scope in itself, i.e. a study based on empirical data from 53 countries, covering not only the business environment, but also issues such as education and religion, can be labeled a deficit for the purpose of the research at hand. Intending to derive detailed differences between Germany and the USA with regard to the constitution of customer loyalty, subtle differences and implications are not sufficiently pronounced in Hofstede's framework. For this reason, a third study, conducted by Hall and Hall (1990), is presented. By restricting their research to an analysis of the USA, Germany, and France only, they are able to draw more in-depth conclusions than Hofstede (2001) and Trompenaars and Hampden-Turner (1997) and their results can be used to complement these broader frameworks.

3.4.3 Hall's and Hall's framework of cultural differences

While Hofstede was the first researcher to conduct a truly large-scale empirical analysis of cultural difference, a somewhat similar concept had already been provided by anthropologist Hall (1960), who defines culture to be a program for behavior (Hall 1976, p. 2). With their book published in 1990, Hall and Hall (1990, p. 4) "offer some conceptual tools to help [their] readers decipher the complex, unspoken rules of each culture." By conducting a qualitative empirical study in the USA, Germany, and France (180 interviews in total), the primary goal of Hall and Hall (1990, p. xiii) is "to help American businesspeople understand German and French psychology and behavior". By dedicating a whole book to cultural differences between US-Americans, Germans and the French, Hall and Hall (1990) close the textual and methodological gap left by Hofstede (2001), who, because of the breadth of his study, was not able go into the subtle details of cultural differences between Germany and the USA.

Since Hall's and Hall's book is mainly addressed at business people, it lacks information on the exact methods used for data gathering and analysis. The information given is limited to the following: "In preparing this study, we conducted in-depth, open-ended interviews with a carefully selected sample of individuals from business and the professions. We also interviewed writers, artists, and educators. Our firsthand observations and these lengthy interviews, coupled with sophisticated models of culture, have enabled us to identify some of the major cultural patterns that business executives need to keep in mind when dealing with their counterparts in other countries." (Hall and Hall 1990, p. xix)

As a result of their interview series, Hall and Hall (1990) identified several key concepts to categorize cultural peculiarities, of which the most important ones, 'context', 'time', 'space', and 'information flow' will be briefly outlined below. In addition, influential differences between Germany and the USA will be pointed out for each concept.

3.4.3.1 High-context and low-context

Devoting special attention to issues of intercultural communication, the concept of high- and low-context communication is the one Hall and Hall (1990) are widely renowned for. Context, in this respect, is the information surrounding an event, and high- and low-context communication are defined by Hall (1976, p. 91): "A high context (HC) communication or message is one in which most of the information is already in the person, while very little is in the coded, explicit, transmitted part of the message. A low context (LC) communication is just the opposite; i.e., the mass of the information is vested in the explicit code." A characteristic feature of LC cultures is the "lack of extensive, well-developed information networks" (Hall and Hall (1990, p. 8), which are a prerequisite of HC cultures. Since members of LC strictly compartmentalize their lives (e.g. personal relationships, work), they require detailed background information every time they act (Hall and Hall 1990, p. 7).

Hall and Hall (1990, p. 8) characterize both Germans and Americans as LC cultures – while Americans only dispose of moderately dense information networks, Germans' information networks are even a little less dense (Hall and Hall 1984, pp. 38-39). However, due to their general similarity with both cultures widely relying on explicitly transmitted information, differences between Germany and the USA cannot be expected on the basis of the communication context.

3.4.3.2 Time

While several different time systems exist, monochronic and polychronic time systems are most important in international business (Hall and Hall 1990, p. 13). In monochronic systems, time is sequential and one action or event clearly follows after the other. In polychronic systems, on the other hand, actions and events overlap and everything happens simultaneously. For this reason, monochronic time is sometimes compared to money, which "can be 'spent', 'saved', 'wasted', and 'lost'." (Hall and Hall 1990, p. 13) An effect of monochronic time systems is the need to prioritize issues, because they cannot be handled simultaneously, leading to an understanding of time as a personal space "which some people are allowed to enter, while others are excluded." (Hall and Hall 1990, p. 14)

According to Hall and Hall (1990), both Germany and the USA adhere strictly to monochronic time systems, which can probably be attributed to the period of the industrial revolution, in which life started to become dictated by "whistles and bells counting off the hours." (Hall and Hall 1990, p. 13) While mainly similar, subtle differences between Americans and Germans exist. German time and consciousness are fixated to the future and Germans are generally "not preoccupied with immediate results" (Hall and Hall 1990, p. 37) and prefer planning for the long-term. Americans on the contrary, are fixated to the present and demand immediate results (Hall and Hall 1990, p. 141), which is in line with Americans' preference for short-term planning. Consequently, it can be expected that US-Americans favor the realization of immediate economic benefits, while Germans prefer arrangements that promise rewards in the long-term.

3.4.3.3 Space

A core system inherent in all cultures is space, which refers to the invisible boundaries between individuals. In contrast to visible boundaries, such as a person's skin, Hall and Hall (1990, p. 10) distinguish between territoriality on the one hand and personal space on the other hand. While the concept of territoriality dates back to the very early times of mankind, when territories were beginning to be claimed and defended, personal space is an invisible bubble of space around each person, "which expands and contracts depending on a number of things: the relationship to the people nearby, the person's emotional state, cultural background, and the activity being performed." (Hall and Hall 1990, p. 11) Territoriality is highly developed in Germany and the USA, although Germans tend to be even more territorial than Americans and extend a feeling of territoriality "to all possessions, including the automobile." (Hall and Hall 1990, p. 10). Analogously, personal space is equally well developed in Germans and Americans of northern European decent and only few people are allowed to penetrate the invisible bubble.

This clearly has implications for relationships. Germans value frankness, honesty, and directness and according to Hall and Hall (1990, p.54), relationships in Germany may go a lot deeper than the comparatively superficial relationships in the USA. However, since "space is sacred" (Hall and Hall 1990, p. 38) to Germans, it is a lot more difficult to form a relationship than it is with Americans. However, relationships in the USA are also more endangered, as "the American's first loyalty is still to self, family, and the career, not the company. [...] Americans also have little personal loyalty in their business dealings. Being pragmatic, they do business where they 'get the best deal', which usually means the best price." (Hall and Hall 1990, pp. 152-153)

3.4.3.4 Information flow

Information flow refers to "how long it takes a message intended to produce an action to travel from one part of an organization to another and for that message to release the desired response" (Hall and Hall 1990, p. 22) and is very slow in low-context or monochronic countries, where "information is highly focused, compartmentalized, and controlled" and "moves almost as if it had a life of its own" (Hall and Hall 1990, p. 23) in high-context or polychronic countries. It was noted before that Germany and the USA are both low-context and monochronic countries, wherefore they are also characterized by restricted information flows. However, it was also shown that Germany is more extreme in both aforementioned issues and information flow can therefore be expected to be even slower in Germany than in the USA and the action a specific message is designed to trigger will therefore be subject to an even greater delay in Germany than in the USA.

3.4.3.5 Summary and critical assessment

According to the above introduced concepts of context, space, time, and information flow, Germany and the USA show more similarities than differences. This does not come as a surprise, however, as Hall and Hall (1990) explicitly focus on Americans of northern European ancestry. Although methodologically beclouded, Hall and Hall (1990) are indeed able to point out some subtle differences between Germans and Americans. These can be attributed to diverging time fixations, where Germans are fixated to the future, while Americans are fixated to the present, striving to achieve immediate economic gratification and to different space concepts that influence relationships and loyalty.

The main point of criticism regarding Hall's and Hall's work is concerned with their choice of interviewees. Both Morain (1991) and Jameson (1992) indicate that only a small and distinct group of individuals from the respective cultures are represented in Hall's and Hall's sample. In fact, Hall and Hall (1990, p. 34) explicitly state that they only included Germans that "inhabit industrialized areas of the north" and Americans "whose forebears came to the U.S. from northern Europe" (Hall and Hall 1990, p. 137). Also, Hall and Hall (1990, p. 139) quote a Gallup survey that showed that "65 percent Fortune 500 executives as well as 68 percent of small-business executives were in this category", thereby acknowledging their restriction, but at the same time claiming it to be textually justified.

While the criticism pointed out above definitely justifies careful handling of Hall's and Hall's findings, it is still appropriate for use in the context of this study, as comprehensive conclusions are not drawn. Rather, Hall and Hall (1990) accentuate some of the findings presented by Hofstede (2001) and results will only be used to complement and engross Hofstede's results.

3.4.4 Trompenaars' and Hampden-Turner's framework

As a third framework, the research by Trompenaars and Hampden-Turner (1997) is broached in this section. Their framework melds aspects of both Hofstede (2001) and Hall and Hall (1990), but is based on a more recent data set. Incorporating 15 years of research, they ground their study both on cross-cultural training programs (over 1,000 conducted in over 20 countries) and on ca. 30,000 questionnaires from 30 companies in 50 countries. "Approximately 75% of participants belong to management (managers in operations, marketing, sales and so on), while the remaining 25% were general administrative staff (typists, stenographers, secretaries)." (Trompenaars and Hampden-Turner 1997, pp. 1-2) In order to provide statistically firm analyses, only samples with "a minimum of 100 people with similar backgrounds and occupations were taken in each of the countries" (Trompenaars and Hampden-Turner 1997, p. 1).

On this basis, Trompenaars and Hampden-Turner (1997) validate seven dimensions in which cultural differences occur: universalism vs. particularism, individualism vs. communitarianism, affective vs. neutral, specific vs. diffuse, achievement vs. ascription, attitudes to time, and attitudes to the environment. These seven dimensions can be subsumed to three categories, relationships (on the basis of the five relational dimensions by Parsons 1951), time, and environment. These seven dimensions, as well as resulting cultural differences between Germany and the USA, will be outlined in the following section.

3.4.4.1 Dimensions of culture and resulting cultural differences

The *universalism vs. particularism* dimension is concerned with rule obedience. While universalists behave rule-based and universally agreed standards are generally adhered to, particularists "focus on the exceptional nature of present circumstances" (Trompenaars and Hampden-Turner 1997, p. 31) and behavior is in accordance with the specific situation, even if that means disobeying certain rules. Since Trompenaars and Hampden-Turner (1997, pp. 35 and 37) find both Germany and the USA to be clearly universalistic countries, cultural differences cannot be identified on the basis of this dimension.

According to Trompenaars and Hampden-Turner (1997, p. 50), the *individualism vs. communitarianism* dimension describes the relationship between individuals and groups in societies, i.e. the underlying question whether individuals are primarily driven by self-interest (individualism) or whether they place common goals and objectives first (communitarianism). While individualists are shown to rely on their own exertion and resources, communitarians rather unite individuals and society. To Trompenaars and Hampden-Turner (1997, p. 57), the USA can clearly be labeled individualistic and they are expected to demonstrate few communitarian characteristics. Germany, on the contrary, is positioned in the middle between communitarianism and individualism and can therefore be presumed to show traits of both sides.

The greatest influence of individualism vs. communitarianism on business relationships is identified in decision making processes (Trompenaars and Hampden-Turner 1997, pp. 60-61). While communitarians foster group decision making, in which other group members always have to be consulted before taking decisions, individualists prefer independent decisions. For this reason, decision making in communitarian countries is a lengthy process, because consensus has to be reached. Decision making in individualistic countries, on the other hand, is rather quick, but implementation can prove difficult when support for individually taken decisions is lacking. On the other hand, decisions are rather stable in communitarian countries. Influences can also be asserted for organizational structure. Trompenaars and Hampden-Turner (1997, p. 63) state that in individualistic cultures, "[m]embers of organizations enter relationships because it is in their individual interests to do so", while the organization in communitarian countries is "a social context all members share and which gives them meaning and purpose." Hence, individualists are much more likely to terminate relationships, if alternative settings have the potential to be more rewarding, while loyalty to a relationship will be higher in communitarian cultures.

Depending on whether cultures are *affective or neutral*, Trompenaars and Hampden-Turner (1997, p. 69) state that in "relationships between people, reason and emotion both play a role". As a result of societal convention, some cultures are more prone to showing emotion than others, which must not be confused with lacking affectivity. Instead, by showing feelings, i.e. being openly emotional, individuals are seeking a direct emotional response, while a neutral approach evokes an indirect response (Trompenaars and Hampden-Turner 1997, pp. 70-71). Since Trompenaars and Hampden-Turner (1997, p. 70) rate both Germany and the USA as being rather neutral, significant cultural differences with regard to this dimension cannot be expected.

Following Trompenaars and Hampden-Turner (1997, p. 81), the *specific vs. diffuse* dimension is concerned with the way relationships between individuals in a society are formed, i.e. "the degree to which we engage others in specific areas of life and single levels of personality, or diffusely in multiple areas of our lives and at several levels of personality at the same time." The data shows that both Germany and the USA tend to be specific, although some degree of difference exists between the two countries (Trompenaars and Hampden-Turner 1997, p. 88 and 93). While the USA are identified as exceptionally specific, Germany is positioned closer to the average. For this reason, some cultural differences are expected and detailed below.

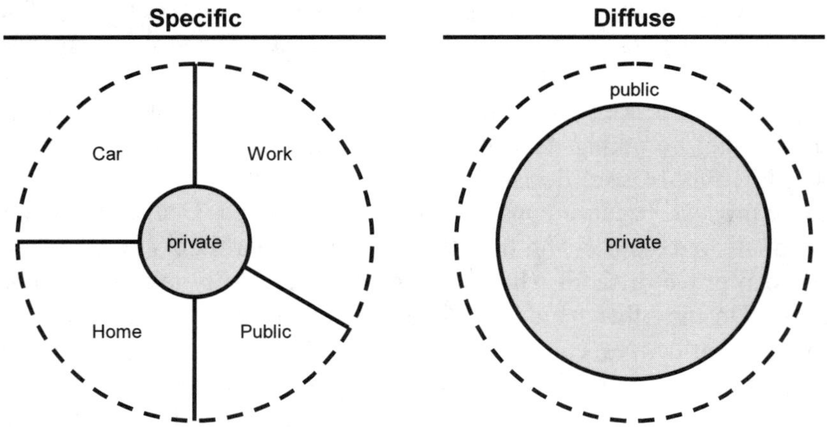

Fig. 3-5. Lewin's model (adapted)

In specific cultures, task relationships prevail, meaning that each encounter or relationship between individuals is clearly separated from all others. On the other extreme, "every life space and every level of personality tends to permeate all others" (Trompenaars and Hampden-Turner 1997,

p. 81) in diffuse cultures. This issue is most precisely reflected in the life spaces model proposed by Lewin (1936), an adaptation of which is shown in Figure 3-5.

In specific cultures, there is a relatively small, private space in the form of the inner circle, which is surrounded by several ‘pie slices’, which stand for different situations in an individual’s life. On the other hand, diffuse cultures are characterized by a relatively large private, inner circle, and an undivided public layer. The main implication is that it is quite easy to engage in a relationship in specific cultures, as the relationship can be limited to one clearly separated space of public life, or, as Trompenaars and Hampden-Turner (1997, p. 82) put it, “being admitted to one public layer is not a very big commitment. You know the other for limited purposes only.” It is more difficult to start a relationship in diffuse cultures, because on the one hand, the public layer is smaller, and on the other hand, once accepted into the inner circle, no more boundaries exist. For this reason, broader relationships are a prerequisite for conducting business in diffuse cultures, while this is not necessary in specific cultures, in which business can be “quick, to the point and efficient.” (Trompenaars and Hampden-Turner 1997, p. 100)

The fifth dimension, *achievement vs. ascription*, refers to they way status is accorded in a culture. In achieved status societies, status is attributed on the basis of achievements, while it is ascribed “to them by virtue of age, class, gender, education” (Trompenaars and Hampden-Turner 1997, p. 102) in others. Although both Germany and the USA belong to the group of countries, where status is accorded on the grounds of achievement, there is a significant gap between them (Trompenaars and Hampden-Turner 1997, pp. 105-106) and resulting cultural differences are pointed out below.

In the USA, where status is accorded on the grounds of achievement, individuals are awarded their rank by merit and will therefore take decisions that foster measurable (usually in monetary terms) performance to secure their advancement to higher status. Since ascribed status managers are not forced to deliver measurable benefits at all times, their orientation may be more longer-term than that of achieved status managers (Trompenaars and Hampden-Turner 1997, p. 108). Consequently, Germans, which more readily ascribe status, may exhibit more long-term orientation in their business conduct than the strictly achievement-driven Americans.

With respect to *attitudes to time*, Trompenaars and Hampden-Turner (1997, p. 121) distinguish between cultures that have a sequential attitude towards time, meaning time is understood as “a series of passing events”, and cultures that have synchronic attitudes to time, such that “past, present and future [are] all interrelated”. As shown by Trompenaars and Hampden-

Turner (1997, p. 128), Germany and the USA have similar time concepts, wherefore significant differences between the two countries cannot be expected.

Regarding *attitudes to the environment*, Trompenaars and Hampden-Turner (1997, p. 141) propose that different cultures assign different roles to themselves and to their natural environments, which is transferred to the question whether individuals are self-determined or not. Asking Germans and Americans about their belief in self-determination, Trompenaars and Hampden-Turner (1997, p. 144) identify a considerable gap. While 66% of German respondents believe to be self-determined, 82% of American respondents do so, indicating that cultural differences may exist, which will be elaborated below.

Depending on their degree of self-determination, cultures can either be classified as inner-directed, meaning they "identify with mechanisms; that is, the organization is conceived of as a machine that obeys the will of its operators", or as outer-directed, tending "to see an organization as itself a product of nature, owing its development to the nutrients in its environment and to a favorable ecological balance." (Trompenaars and Hampden-Turner 1997, p. 141) While individuals in inner-directed cultures see their fates as self-determined, individuals in outer-directed cultures believe their destiny is directed by forces outside their control. The most important effect of the dimension at hand is the difference ascribed to relationships. In outer-directed cultures such as Germany, great effort is invested into maintaining relationships. In inner-directed cultures, on the contrary, it is important to "win your objective" (Trompenaars and Hampden-Turner 1997, p. 155), such that consensus and stability are deemed less important than imminent success.

3.4.4.2 Summary and critical assessment

As shown in the previous section, Trompenaars and Hampden-Turner (1997) were able to identify moderate differences between the USA and Germany in four out of seven cultural dimensions – significant differences were not found with regard to the three dimensions of 'universalism vs. particularism', 'affective vs. neutral', and 'attitudes to time'.

The USA score higher on individualism than Germany, suggesting that interpersonal relationships could be more meaningful to Germans, who also value personal loyalty more highly. In addition, group decision making and consensus are nurtured in Germany. Concerning the "specific vs. diffuse" dimension, the USA are strictly specific, while Germany exhibits a considerable amount of diffuse traits. Therefore, personal factors are concluded to be more important to business relationships in Germany than in the USA, where business conduct is strictly results oriented. With regard to status, both countries tend towards according status on grounds of achievements. However, ascription is still a factor in Germany and consequently, managers here have to be less focused on short-term and visible results and performance. Finally, Germans were found to be less self-determined than Americans, which may make it easier and more important for them to maintain relationships.

The framework of Trompenaars and Hampden-Turner (1997) has been both praised and criticized. Supporters laud the framework's high relevance to the business world (e.g. Mendonsa 1999), as well as its accessible style of presentation, which forgoes "remain[ing] locked in the domain of academic research papers" and "drown[ing] the audience in statistics." (McIntosh 1999, p. 144)

Critics of the research raise some objections, mainly concerning the employed methodology and academic rigor. Gallagher (1996, p. 231), for instance, criticizes the absence of the full questionnaire and states that the work is "no theoretical breakthrough". In fact, it is Hofstede who disapproves the most. To him, the empirical results appear incomplete and randomly chosen, "respondents' answers were not summarized into country scores on the seven 'dimensions'" (Hofstede 1997, p. 287), and it is not completely clear where a country is positioned compared to all others (Hofstede 1996, p. 190).

While the research presented by Trompenaars and Hampden-Turner (1997) is clearly targeted at business practitioners and even though the approach chosen does not fully satisfy scientific standards, it is still valuable to complement the other two presented frameworks. Firstly, it builds upon the works of Hofstede and Hall and Hall and introduces some new aspects and a fresh view. Secondly, it provides more recent data, which is a major advantage over Hofstede's study. Trompenaars and Hampden-Turner (1997) can therefore be applied to validate Hofstede's findings and to enrich the overall picture of cultural differences drawn in this chapter.

3.4.5 Overview of cultural differences

Examining the three studies on cultural differences altogether, ten significant differences between Germany and the USA can be extracted. Eight of them originate from Hofstede's research, and of these, four are supported by Hall and Hall and/or Trompenaars and Hampden-Turner. Only the stronger long-term orientation prevalent in Germans and Americans' focus on achieving immediate economic benefits cannot be backed by

Hofstede's findings and are derived solely on the basis of the other two studies.

A summary of differences is given in Table 3-1. For each difference, the USA are set as point of reference, and Germany's deviation from this point is provided. Also, the provenience of each difference, i.e. if it originates from Hofstede (2001), Hall and Hall (1990), and/or Trompenaars and Hampden-Turner (1997), is stated.

Table 3-1. Summary of cultural differences

In comparison to US-Americans, Germans...	Hofstede	Hall/Hall	Trompenaars/ Hampden-Turner
...extend trust more readily, especially to in-group members.	●		
...are more accepting of information sharing.	●		
...dread change and innovation.	●		
...adhere to rules and regulations more strictly.	●		
...place higher emphasis on stable relationships.	●		●
...are more loyal towards relationships, especially with in-group members.	●	●	
...seek consensus and try to avoid conflict.	●		●
...require personal bonding for maintaining long-term business relationships.	●	●	●
...place lower emphasis on immediate economic benefits.			●
...are rather long-term oriented.		●	●

3.5 Theory synopsis

The study at hand is built on the theoretical fundament of social exchange theory and its extensions, equity theory and commitment trust theory. As pointed out in the respective sections (3.1.2, 3.2.2, 3.3.2), these theories are appropriate for explaining a variety of potential determinants of customer loyalty in the setting scrutinized here, i.e. the long-term relationships typically present between LSPs and their customers. Certainly, other theories could be added to include other facets of customer loyalty. The research by Wallenburg (2004), which this study is based on, employed a broader approach and encompassed exit-voice theory, dissonance theory, risk theory, learning theories, and transaction cost theory in addition to the three theories used here. With the exception of transaction cost theory, the other theories were used to provide additional aspects for conceptualizing the determinants of customer loyalty proposed in the present research and are omitted here with the aim of providing leaner and more focused research. It was shown previously in this chapter, and will be explained in more detail in chapter 4, that social exchange theory, equity theory, and commitment trust theory by themselves provide sufficient grounding for conceptualizing the determinants examined here.

In addition, Wallenburg (2004) used transaction cost theory as a basis for explaining economic barriers to switching, such as switching costs and specific investments. Since these factors were shown not to contribute significantly to the explanation of customer loyalty between LSPs and their customers, they are not studied here and consequently, transaction cost theory is not required.

The combination of social exchange theory, equity theory, and commitment trust theory, however, is useful and necessary to tackle research questions one and two proposed before. The three theories build on a similar conceptual framework, sharing most of their underlying premises. While social exchange theory provides the basic framework of analysis, equity theory offers further detail for explaining the additional phenomenon of 'fairness', and commitment trust theory is explicitly concerned with affective commitment and trust.

The identification of cultural differences between Germany and the USA with regard to the formation of customer loyalty necessitates the inclusion of cultural theory. While social exchange theory, equity theory, and commitment-trust theory offer a general, more or less culturally unbiased perspective on customer loyalty and its drivers, they do not offer insights into cultural distinctions that may occur in different cultural settings. Hence, the complementary cultural models of Hofstede, Hall and Hall, and Trompenaars, as well as the empirical evidence provided in their research, are included to examine research question four.

4 Research model

After having introduced the theories, which will be applied in the present research, this chapter will be devoted to devising a comprehensive explanatory model of customer loyalty, as well as the moderating influences of relationship characteristics and cultural differences between Germany and the USA. For that, the first two sections, 4.1 and 4.2 will conceptualize customer loyalty and its determinants, while the subsequent section derives hypotheses on the relationships between these conceptualized constructs, resulting in a comprehensive model of customer loyalty. Then, section 4.4 will conceptualize several important relationship characteristics and develop hypotheses on their effects on the linkages between customer loyalty and its determinants. Finally, section 4.5 will build upon the previously identified cultural differences between the USA and Germany to posit a set of hypotheses on how these two countries are expected to differ with regard to the constitution of customer loyalty.

4.1 Conceptualization of customer loyalty

In chapter 2.1.2, behavioristic and neo-behavioristic customer loyalty concepts were briefly described. It was shown that behavioristic concepts refer only to observable behavior and therefore disregard the roots of observable behavior. Neo-behavioristic concepts, on the contrary, allude to attitudes and intentions, as well as observable behavior, thereby going to the roots of customer loyalty. Sophisticated conceptualizations of customer loyalty were also shown to be multi-dimensional, including repeat purchases, additional purchases and referrals.

In the service loyalty context, however, Wallenburg (2004, pp. 16-17) identifies problems with respect to the validity of behavioristic loyalty measurement. Especially between LSPs and their customers, the long duration of contractual relationships impedes the observation of behavioristic loyalty. In contrast to consumer goods, buyers of logistical services do not make frequent purchase decisions, but commit themselves contractually to their LSPs for relatively long periods of time. Consequently, switches from one provider to another typically take place at the end of contracts and retention in current relationships between LSPs and their customers is not determined by customer loyalty, but is created by contracts. For this reason, a neo-behavioristic view is usually employed, which would ideally allude to attitudes, intentions, and observable behavior. It was shown in chapter 2.1.2.2, however, that it is not advisable to include attitudes, as these do not tend to 'involuntary' loyalty, such as that resulting from economic constraints.

For an ideal assessment of customer loyalty, intentions would therefore have to be assessed and aligned to actual behavior. As a result of the long contractual commitments between LSPs and their customers, however, surveys would have to be conducted at two different points in time: a first survey that determines intentions in existing relationships, and a second survey at a later time, in which actual behavior is identified and compared to the previously surveyed intentions (Wallenburg 2004, pp. 23-24). But because the second survey would have to be conducted with a great delay relative to the first survey, two major problems would arise: On the one hand, longitudinal surveys over a time horizon of several years are nearly impossible to conduct, as respondents would mostly be in different positions, sometimes even with other employers. On the other hand, exogenous factors, such as the respondent's firm's competitive environment, may well change in between surveys, and therefore inhibit an unambiguous comparison of intentions and actual behavior. For these reasons, actual purchasing behavior is not surveyed in this study. As in most other studies on customer loyalty in long-term relationships (e.g. Bloemer and de Ruyter 1999, Stank, Goldsby, and Vickery 1999, Caruana 2000, Sirdeshmukh, Singh, and Sabol 2002, Stank et al. 2003, Wallenburg 2004, Knemeyer and Murphy 2005), customer loyalty will be assessed by surveying intentions only.

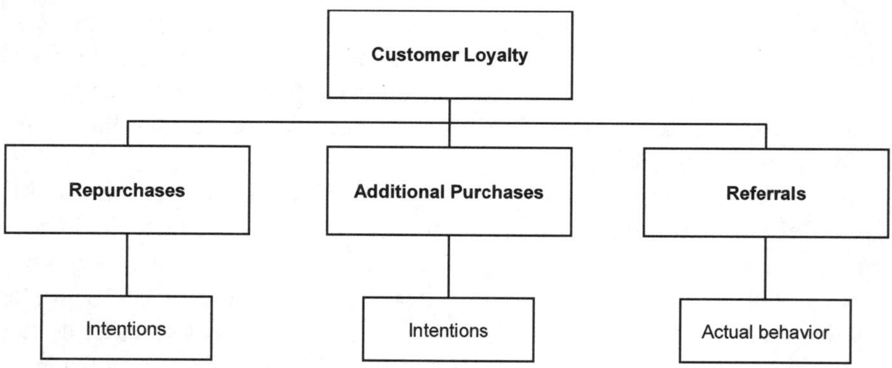

Fig. 4-1. Conceptualization of customer loyalty used in this study

In contrast to purchasing behavior, referrals can be made at any time, independent from contract duration. For this reason, actual referral behavior can be surveyed. Since relationships between buyers and providers of logistical services are typically characterized by a high degree of stability, referral intentions do not have to be surveyed, because they are not expected to differ from actual referral behavior.

In this way, the conceptualization of customer loyalty is analogous to the one proposed and validated by Wallenburg (2004, pp. 59-61), and summarized in Figure 4-1. Repurchases of the same service (portfolio) from the same LSP, as well as additional purchases of services exceeding the services currently rendered by an LSP are assessed by surveying intentions, while referrals of the LSP by its customers are captured by surveying actual referral behavior.

4.2 Conceptualization of customer loyalty determinants

In this section, determinants of customer loyalty, as proposed in chapters 3.1.2, 3.2.2, and 3.3.2, will be conceptualized. These are 'satisfaction', 'proactive improvement', 'fairness', 'commitment', 'trust', and 'alternatives'. Figure 4-2 gives an overview of the determinants discussed in this chapter. Since all of these factors were already part of the research of Wallenburg (2004), conceptualizations will deliberately be kept short.

4.2.1 Satisfaction

In contrast to other psychological marketing constructs, satisfaction is also used in every-day language and associated with a positive state of mind (Wallenburg 2004, p. 93). Such a vague understanding of the notion of satisfaction of course does not suffice for its application in research. Thus, researchers engaged in the study of satisfaction provide a multitude of concepts that try to substantiate satisfaction, but because of the great variety of research in this field, a common understanding has not been established. Conceptualizations of satisfaction differ especially with regard to the type and number of components to be included.[1] While researchers agree that satisfaction is a "postchoice evaluative judgment" (Westbrook and Oliver 1991, p. 84) resulting from a mental comparison (Oliver 1997, p. 12), there is no unison on the object underlying satisfaction. Oftentimes, transactions within relationships are supposed to be responsible for satisfaction. This,

[1] See Giese and Cote 2000 for an overview of conceptualizations of satisfaction.

however, is due to the circumstance that most research on satisfaction focuses on consumer goods. It is questionable, whether this approach is applicable for defining satisfaction in a business-to-business service context marked by long-term cooperation, as it is prevalent between LSPs and their customers. In long-term relationships, satisfaction is not only a result of individual transactions. Instead, such relationships facilitate the formation of relational governance structures and, as posited by social exchange theory, immediate benefits may be foregone for future benefits. If an exchange partner is therefore confident that the entirety of transactions is beneficial, momentary dissatisfaction may loose weight in the overall judgment (Homburg, Giering, and Menon 2003, p. 40).

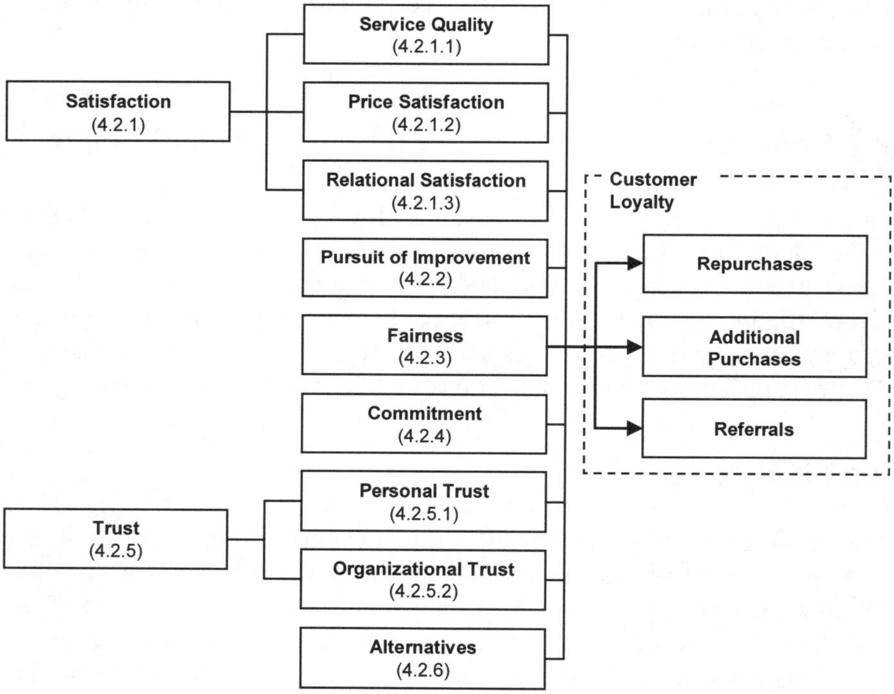

Fig. 4-2. Determinants of customer loyalty

Of great interest to researchers of satisfaction is its formation. While a multitude of approaches appear suitable for conceptualizing the formation of satisfaction, the majority of researchers reverts to the disconfirmation paradigm (Churchill and Suprenant 1982, p. 491, Cadotte, Woodruff, and Jenkins 1987, p. 305, Brady, Cronin, and Brand 2002, p. 17). "If perceived performance exceeds a consumer's expectations (a positive disconfirmation), then the consumer is satisfied. But if perceived performance falls short of his or her expectations (a negative disconfirmation), then the consumer is dissatisfied." (Spreng, MacKenzie, and Olshavsky 1996, p. 15) With this in mind, satisfaction in the context of this study will be understood as the cognitive and affective evaluation of the entirety of experiences made with an LSP (Wallenburg 2004, pp. 95-96).

Trying to examine satisfaction in the context of the relationships between LSPs and their customers, however, Wallenburg (2004, p. 264) shows that a global measure of satisfaction is too general. Instead, he concludes that different dimensions of satisfaction have to be considered in order to be able to identify individual aspects constituting satisfaction. For this reason, three dimensions of satisfaction, conceptually similar to the logistics performance dimensions proposed by Stank et al. (2003, pp. 28-29), will be conceptualized. These three dimensions, satisfaction with service quality, price, and relational performance, are introduced below.

4.2.1.1 Service quality

Examining the extant body of customer satisfaction research, performance based satisfaction appears to be a frequently studied subject (e.g. Bitner 1990, Babakus and Boller 1992, Boulding et al. 1993, Taylor and Baker 1994, Zeithaml, Berry, and Parasuraman 1996, Sirohi, McLaughlin, and Wittink 1998, van Dyke, Prybutok, and Kappelmann 1999), especially with regard to service performance perceived by consumers. Like other dimensions of satisfaction, satisfaction with service performance develops as a result of positive disconfirmation, i.e. when the achieved service level exceeds expectations.

Recurring to service satisfaction in the classic sense, however, poses problems in the three dimensional treatment of satisfaction used in this study. While "the dimensions underlying quality are fairly specific[,...] satisfaction judgments have a broader range of dimensions that also include quality aspects" (Caruana, Money, and Berthon 2000, p. 1338). Also, "perceived service quality is a global judgment" (Parasuraman, Zeithaml, and Berry 1988, p. 16), while "satisfaction assessments require customer experience" (Caruana, Money, and Berthon 2000, p. 1338). In fact, Cronin Jr. and Taylor (1992) show that quality is an antecedent of satisfaction. Due to the specific context of this study, in which the evaluation of service performance is to be isolated from the other dimensions of satisfaction (relational satisfaction, price satisfaction), and because the evaluation of quality in relationships between LSPs and their customers is not only based on a global judgment, but also on specific experiences, service quality is measured instead of satisfaction.

Corresponding to the conceptualization developed by Wallenburg (2004, p. 98), service quality in this research is defined as the cognitive and affective evaluation of all experiences made by a customer with the services provided by an LSP. Hence, service quality is assessed according to the disconfirmation paradigm (Brady and Cronin 2001, p. 35) and it is independent from the absolute and objective service level provided by the LSP. Rather, service quality as perceived by the customer is considered, as only this leads to satisfaction.

4.2.1.2 Price satisfaction

"Satisfaction literature offers little insight into the effect [...] pricing decisions might have on customer satisfaction." (Voss, Parasuraman, and Grewal 1998, p. 46) Wallenburg (2004, p. 98) points out that typical conceptualizations of customer satisfaction, only include price, in conjunction with other factors, to determine the expectation level. This is due to the perception that purchasing decisions are made that comprise all relevant factors including price, to then form expectations, and ultimately evaluate the received service against the expectation level. Implicitly, this approach assumes that the actual service level has to be maximized to create satisfaction and expectations are treated as given.

It is questionable, however, whether this view is appropriate in all contexts. In fact, Varki and Colgate (2001) argue that price is an important driver of customers' perceptions of value, which in turn influence satisfaction. Also, the importance of price for customer satisfaction has been validated in empirical studies (e.g. Voss, Parasuraman, and Grewal 1998, Lapierre, Filiatrault, and Chebat 1999).

Price can influence satisfaction in several ways, as shown by Wallenburg (2004, p. 99):

- Logistics outsourcing decisions are often not made by one person alone; decision memos are drafted and the final decision is made by a manager in a higher hierarchical position. In this case, the subordinate has to justify its proposal which can lead to a stronger emphasis on cost or price aspects.
- Monetary measures are generally the most important criteria in corporate decision making. Since service performance is harder to quantify than cost or price, and consequently also harder to compare to market alternatives, price plays an important role in practice.
- Sometimes, additional service performance does not create additional utility for the customer. Hence, satisfaction is the greater, the lower the price for a given level of service performance is.
- In long-term contractual relationships, as they are common between LSPs and their customers, dissatisfaction may arise, even if the price of

the contracted service was initially perceived as adequate. This can occur, when alternative LSPs offer comparable services at lower prices and can lead to dissatisfaction that is not created by the quality of service, but by its price.

Overall, price can be assumed to have a decisive effect on the creation of satisfaction in the industrial service context. In accordance with the definition of service quality given above and the conceptualization provided by Wallenburg (2004, p. 99), price satisfaction in this study is based on the disconfirmation paradigm and is defined as the affective and cognitive evaluation of all experiences made by a customer with the prices offered by an LSP for the contracted logistical services.

4.2.1.3 Relational satisfaction

Research on customer satisfaction does not deny the relevance of relational factors within long-term relationships. However, little work has been done that stresses the link between good relationship interaction and satisfaction. Some empirical evidence is given by Stank, Goldsby, and Vickery (1999), who indicate that there actually is a strong link between relational performance and satisfaction. Other research, e.g. by Bauer (2000), pursues a very narrow approach, by exclusively examining effects of relational factors on service performance. By doing this, social aspects as well as potential cost reductions are neglected.

In order to grasp all relevant facets of satisfaction, relational satisfaction is used in this study to complement service quality and price satisfaction. Following the definition of the latter two and Wallenburg's (2004, p. 100) conceptualization, relational satisfaction is defined as the affective and cognitive evaluation of all experiences made by a customer with the interaction with and relationship to an LSP.

4.2.2 Proactive improvement

As delineated before, the changing market environment and increasing service requirements place new challenges on LSPs. Not only do they have to operate existing logistics systems, but they also have to improve in order to become more efficient and effective, e.g. because many firms today require LSPs to pass on efficiency gains in the form of price decreases at prescribed times during the relationship. In order to be able to improve, LSPs have to be innovative and must be willing to incorporate their innovative abilities in the particular relationship with the customer. These thoughts lead to the development of the concept of proactive improvement by Engelbrecht (2004) and Wallenburg (2004).

Even though researchers acknowledge the importance of continuous improvements in long-term relationships in general (e.g. Narayandas and Rangan 2004, p. 74) and in logistics outsourcing relationships in particular (e.g. Boyson et al. 1999, p. 95), the concept of proactive improvement so far has only been studied by Engelbrecht (2004) and Wallenburg (2004). Therefore, very little indication is given on how to conceptualize proactive improvement. The reason why proactive improvement has hardly been treated in marketing literature, despite of its obvious appeal, is probably attributable to the circumstance that it is only important in long-term relationships. However, as pointed out before, the majority of the marketing literature focuses on consumer goods and retailing.

Conceptually, proactive improvement can be based within the framework of social exchange theory and will have two major effects. On the one hand, proactive improvement will lead to performance meliorations, as logistics systems are continuously improved. Going back to the CL/CLalt-concept introduced in chapter 3.1.1, a customer will achieve higher benefits (in the form of better performance) from an LSP that displays proactive improvement. This will lead to a comparative advantage for the LSP, when the customer compares the achieved benefit to CLalt.

On the other hand, proactive improvement creates a positive signaling effect. When an LSP conveys that it proactively improves logistics systems, this will create confidence with the customer. In consequence, commitment and trust are nurtured, relational governance norms are established, and customers might be willing to forego present benefits for future benefits, when the LSP displays proactive improvement. In addition, Lemon, White, and Winer (2002, p. 12-13) point out that customers are generally forward looking, such that the positive signaling effect in itself should foster a customer's propensity to continue a relationship with an LSP.

Empirically, two works so far studied proactive improvement. In the customer loyalty context, Wallenburg (2004) was able to validate significant effects on service quality, additional purchases, and referrals. In a study that examined success factors of logistics outsourcing, Engelbrecht (2004) found that proactive improvement significantly increases goal achievement in logistics outsourcing projects.

In the sense of this study, proactive improvement is therefore defined as a customer's affective and cognitive evaluation of an LSP's efforts to proactively improve logistics systems.

4.2.3 Fairness

As outlined in chapter 3.2, fairness is important in buyer-seller relationships, because if individuals perceive a relationship to be unfair, severe behavioral consequences may result. The assessment of fairness is a comparative judgment that weighs the outcome, i.e. an input-output-ratio, of one party against the other's (Xia, Monroe, and Cox 2004, p. 1). Fairness is induced if both parties' outcomes are balanced. In this sense, fairness is based on the concept of reciprocity, which Neuert et al. (2005, pp. 340-341) ostensively compare to the proverb "tit-for-tat".

An objective assessment of fairness would require perfect information about the other party's input-output-ratio. Since this requirement is typically not met as a result of imperfect information and limited information processing capabilities, Xia, Monroe, and Cox (2004, p. 2) argue that the judgment of fairness is subjective. Also, they "propose that affect is an important element", because "[a] buyer may have feelings of unease or guilt when the inequality is to his or her advantage but feelings of anger or outrage when the inequality is to his or her disadvantage." (Xia, Monroe, and Cox 2004, p. 2) Accordingly, fairness is induced when individuals perceive both parties' outcomes as equitable, including both economic and social aspects in the judgment.

In this study, fairness is therefore defined as a customer's affective and cognitive evaluation of the balance between its own input-output-ratio and the LSP's estimated input-output-ratio.

4.2.4 Commitment

According to Gundlach, Achrol, and Mentzer (1995, p. 1), "[commitment] is an essential part of successful long-term relationships" and has been defined as "an implicit or explicit pledge of relational continuity between exchange partners." (Dwyer, Schurr, and Oh 1987, p. 19) Anderson and Weitz (1992) note that committed partners are willing to make relationship specific investments, because they are confident about the long-term potential of the relationship.

The most commonly cited definition of commitment goes back to Moorman, Zaltman, and Deshpandé (1992, p. 316): "Commitment to the relationship is defined as an enduring desire to maintain a valued relationship." Building on this broad understanding of commitment, Morgan and Hunt (1994) provide a more specific definition of commitment that incorporates research in the fields of social exchange (Cook and Emerson 1978), marriage (Thompson and Spanier 1983), and organizations (Meyer and Allen 1984). According to Morgan and Hunt (1994, p. 23), relationship commitment can be defined "as an exchange partner believing that an ongoing relationship with another is so important as to warrant maximum efforts at maintaining it; that is, the committed party believes the relationship is worth working on to ensure that it endures indefinitely."

While Pritchard, Havitz, and Howard (1999) state that work on commitment began in the sociology and psychology disciplines, several researchers have already investigated the role of commitment in buyer-seller relationships (e.g. Dwyer, Schurr, and Oh 1987, Anderson and Weitz 1992, Moorman, Zaltman, and Deshpandé 1992, Anderson, Hakansson, and Johanson 1994, Morgan and Hunt 1994, Gundlach, Achrol, and Mentzer 1995, Hocutt 1998, Moore 1998, Rodriguez and Wilson 2002, Wong and Sohal 2002, Knemeyer, Corsi, and Murphy 2003, Wallenburg 2004).

So far, the majority of researchers employed a uni-dimensional approach to commitment. However, this approach has been subject to criticism by researchers such as Meyer and Allen (1984) and McGee and Ford (1987), who postulate that affective and cognitive commitment should be regarded as two distinct dimensions. Allen and Meyer (1990) even add normative commitment as a third dimension. At the core of their proposition lies the assumption that firms either remain in relationships because they want to or because they are obligated to. While this multi-dimensional approach has been used in a few recent empirical studies (e.g. Wetzels, de Ruyter, and van Birgelen 1998, Gruen, Summers, and Acito 2000), Wallenburg (2004) showed a significant effect of affective commitment on customer loyalty in relationships between LSPs and their customers, but was not able to validate effects of normative and cognitive commitment. In order to accomplish parsimony, this study will employ a uni-dimensional approach following Verhoef (2003, p. 31) and solely "focus on the affective component of commitment, that is, the psychological attachment, based on loyalty and affiliation, of one exchange partner to the other." As such, commitment here is defined in line with Wallenburg (2004, pp. 110-112) as a customer's affective evaluation of its psychological attachment to an LSP.

4.2.5 Trust

According to Arrow (1975, p. 24), trust is essential in all interorganizational relationships and enables companies "to focus on the long-term benefits of the relationship" (Doney and Cannon 1997, p. 35), which is especially important in markets characterized by increasing competition. In order to meet competition more flexibly, many firms build collaborative relationships with buyers and suppliers. In the context of professional services, however, Crutchfield (2001, p. 19) points out that buyers "face considerable uncertainty due to service characteristics such as intangibility, complexity, and long-term horizon of delivery." Trust can counteract the strong potential for service failure and negative outcomes and thus reduces the perception of risk (Lane and Bachmann 1996, p. 390). As a result of trust, customers feel safe and secure in the relationship and will, in consequence, commit themselves to the continuation of the relationship (Liljander and Strandvik 1995, Doney, Cannon, and Mullen 1998).

While there is no general unity about the definition of trust in the buyer-seller-context, an often cited definition is again provided by Moorman, Zaltman, and Deshpandé (1992, p. 315), who define trust "as a willingness to rely on an exchange partner in whom one has confidence." This definition explicitly includes the behavioral intention to act, i.e. to accept risks that would not be accepted if a partner was not trusted (Morgan and Hunt 1994, p. 23) and Moorman, Deshpandé, and Zaltman (1993, p. 82) "argue that both belief and behavioral intention components must be present for trust to exist." Also, "[r]isk is a necessary condition for trust" (Molm, Takahashi, and Petersen 2000, p. 1422) and as an antecedent to risk, Moorman, Deshpandé, and Zaltman (1993, p. 82) regard vulnerability as another necessary condition, because without it, "outcomes are inconsequential for the trustor".

Another implication of this definition is its inclusion of two distinct components of trust, i.e. credibility and benevolence. Ganesan (1994, p. 3) notes that credibility "is based on the extent to which the retailer believes that the vendor has the required expertise to perform the job effectively and reliably", while benevolence "is based on the extent to which the retailer believes that the vendor has intentions and motives beneficial to the retailer when new conditions arise, conditions for which a commitment was not made." This also marks the duality of trust. On the one hand, trust is a result of credible and benevolent behavior by the supplier, while on the other hand, it connotes the expectation that behavior will remain credible and benevolent in the future. Thus, credibility and benevolence are antecedents and consequences of trust at the same time.

Lane (2000, p. 3) summarizes that most concepts of trust share the following three elements:

- Interdependence between trustor and trustee. Trust is only relevant, if "one's consequential activities depend on the prior action or co-operation of another person" (Lane 2000, p. 3; see also Luhmann 1973, p. 43).

- Risk or uncertainty in exchange relationships can be counteracted by trust. Parties may be exposed to opportunistic behavior as a consequence of uncertainty and risk and the resulting dangers can be limited by the existence of trust (Yamagishi, Cook, and Watabe 1998, p. 170).
- Expectation that vulnerability is not taken advantage of. Accepting risks may lead to vulnerability and the existence of trust reduces the involved dangers.

Hence, trust "reduces the perception of risk associated with opportunistic behaviors of the vendor [...], it increases the confidence of the retailer that short-term inequities will be resolved over a long period, and [...] it reduces the transaction costs in an exchange relationship." (Ganesan 1994, p. 3) In the long-term, trust and the associated risk reduction bind buyers to their suppliers, as noted by Crutchfield (2001, p. 20).

Now that the notion of trust and its importance have been clarified, a definition must be provided as to who trustors and trustees are. There is little doubt that trustors can only be individuals. Trustees, however, can be both organizations and individuals within organizations (Doney and Cannon 1997). Individuals can be trusted, as they are the key interface to an organization. In addition, organizations as a whole can be trusted (Sako 2000, Sydow 2000). For instance, postal services are trusted to reliably deliver mail, independent of trust in a specific person, e.g. the mail man.

Thus, this study will use the above stated definition of trust established by Moorman, Zaltman, and Deshpandé (1992) and, like Wallenburg (2004, pp. 104-108), distinguish between a customer's trust in an individual, i.e. personal trust, and trust in an organization as a whole, i.e. organizational trust.

4.2.6 Alternatives

In the overview of social exchange theory given in chapter 3.1, alternatives to the existing provider were shown to be important, as they are used as a comparison standard (CL_{alt}) to determine if a relationship should be continued. Also, non-existance of customer loyalty, i.e. switching to another provider, is only possible, if alternative providers exist. Despite the obvious importance of market alternatives for customer loyalty, few researchers so far have included them in their models. While some researchers studied switching behavior, alternatives only played a subordinate role (e.g. Keaveney 1995). Other researchers incorporated alternatives, but did not include them in a comprehensive model of customer loyalty determinants (e.g. Sriram and Mummalaneni 1990, Ping 1993, Bolton, Kannan, and Bramlett 2000).

In his study on customer loyalty determinants, Wallenburg (2004) did indeed find out that the availability and quality of alternatives are important in the customer loyalty context. Therefore, alternatives will also be incorporated in this study and are defined as a customer's cognitive and affective evaluation of the availability of adequate alternative LSPs in the market-place.

4.3 Comprehensive model of customer loyalty

The present section builds upon the conceptualizations of customer loyalty and its determinants in the preceding section and especially on the customer loyalty model conceptualized by Wallenburg (2004).

In order to develop a model of customer loyalty, a set of hypotheses is proposed, which state all expected effects of determinants on loyalty, as well as interdependencies between the determinants. Again, argumentations are kept short, as the majority of relationships were already included in the model developed by Wallenburg (2004).

4.3.1 Hypotheses for customer loyalty determinants

4.3.1.1 The effects of "service quality"

The body of literature on satisfaction leaves no doubt that overall satisfaction is important for customer loyalty (Innis and La Londe 1994, p. 2, Sureshchandar, Rajendran, and Anantharaman 2002, p. 364). As shown in chapter 4.2.1.1, service quality is one of the three facets of satisfaction scrutinized in this study. According to social exchange theory, the benefit and consequently the satisfaction perceived by the customer depend both on cost and on performance aspects, the latter being represented by service quality.

In addition, several authors have already examined the relationship between service quality and customer loyalty (see Table 2-1). Therefore, the following hypothesis regarding the effect of service quality on customer loyalty can be formulated:

$H_{D1a\text{-}c}$: Service quality has a direct positive effect on (a) repurchases, (b) additional purchases, and (c) referrals.

The details on commitment presented in chapter 4.2.4 suggest that it requires a "valued relationship" (Moorman, Zaltman, and Deshpandé 1992, p. 316), which partners believe "is worth working on to ensure that it endures indefinitely." (Morgan and Hunt 1994, p. 23). It is quite clear that for a relationship to be valued, service quality must exist. In fact, Wetzels, de Ruyter, and van Birgelen (1998) were able to validate the general link between satisfaction and commitment and Wallenburg (2004) even showed a positive effect of service quality on commitment. Therefore, the following hypothesis is proposed:

H_{D2}: Service quality has a direct positive effect on commitment.

As stated in the corresponding section, trust is built on the expectation that the other party is benevolent. Therefore, trust is increased in long-term relationships, if the other party demonstrates, through actions, that it is trustworthy. Such an action would be the provision of satisfactory service quality, which may express trustworthiness, both of an organization and of an individual. As proposed by Wallenburg (2004, pp. 219-220), the following hypothesis is stated:

$H_{D3a\text{-}b}$: Service quality has a direct positive effect on (a) personal trust and (b) organizational trust.

Service quality also has positive effects on the other two dimensions of satisfaction: price satisfaction and relational satisfaction. According to the disconfirmation paradigm, satisfaction is achieved if initial expectations about the service are positively disconfirmed. Price satisfaction can be increased if either price is lower than originally expected or if service quality is higher. Therefore, higher service quality should lead to greater satisfaction with price, which is expressed in the following hypothesis:

H_{D4}: Service quality has a direct positive effect on price satisfaction.

Even though disagreements are inevitable in any relationship (Anderson and Narus 1990, p. 56), a high extent of service quality provided by an LSP can be expected to reduce the level of friction, such that the relationship should be perceived as more rewarding by the customer. This linkage was shown to be particularly strong and highly significant by Wallenburg (2004, p. 235) and the following hypothesis is therefore formulated:

H_{D5}: Service quality has a direct positive effect on relational satisfaction.

As stated in chapter 4.2.3, fairness is based on the concept of reciprocity and induced if both parties' outcomes are balanced. Outcomes are defined as the input-output-ratios of both parties and outputs are highly dependent on the perceived service quality. Even though Wallenburg (2004, p. 235) was not able to provide empirical evidence for this relationship, the following hypothesis is stated to once more scrutinize the link:

H_{D6}: Service quality has a direct positive effect on fairness.

4.3.1.2 The effects of "price satisfaction"

The previous section showed that satisfaction increases customer loyalty and that satisfaction consists of cost and performance elements. Hence, since the cost of an outsourced service is strongly influenced by the price paid for it, price satisfaction is surmised to increase customer loyalty. Wallenburg (2004, pp. 100-102) conceptualized the loyalty-effects of price satisfaction and found partial empirical support for it and Ranaweera and Neely (2003, p. 238) located a strong correlation with customer retention, such that the following hypothesis is proposed:

$H_{D7a\text{-}c}$: Price satisfaction has a direct positive effect on (a) repurchases, (b) additional purchases, and (c) referrals.

It was also shown that balanced outcomes lead to perceived fairness. While service quality affects the output side of the equation, price corresponds to the input side (Wallenburg 2004, pp. 218-219). Therefore, higher price satisfaction leads to an improved perception of fairness:

H_{D8}: Price satisfaction has a direct positive effect on fairness.

4.3.1.3 The effects of "relational satisfaction"

In addition to the 'economic' aspects of service quality and price satisfaction, social exchange theory establishes that satisfaction is also created as a result of social aspects. The quality of the relationship, which – if it exceeds the initially expected level – creates satisfaction, will therefore increase customer loyalty (Wallenburg 2004, pp. 103-104). Prior empirical work by De Wulf, Odekerken-Schröder, and Iacobucci (2001) supports the following hypothesis:

$H_{D9a\text{-}c}$: Relational satisfaction has a direct positive effect on (a) repurchases, (b) additional purchases, and (c) referrals.

Analogous to the line of argumentation for the link between service quality and commitment given above, relational satisfaction enhances commitment, because a valued relationship does not only require good service quality, but also satisfaction with the way parties work together, i.e. relational satisfaction (Wallenburg 2004, pp. 220-221). Consequently, the following hypothesis is stated:

H_{D10}: Relational satisfaction has a direct positive effect on commitment.

The development of trust in a relationship is hampered by uncertainty about the other party's intentions and behavior. Well working relationships, reflected in high relational satisfaction, limit behavioral uncertainty and will thus increase trust (Wallenburg 2004, pp. 219-220). Hence, the following hypothesis can be formulated:

$H_{D11a\text{-}b}$: Relational satisfaction has a direct positive effect on (a) personal trust and (b) organizational trust.

As do the other two facets of satisfaction, relational satisfaction increases perceived fairness (Wallenburg 2004, pp. 218-219), leading to the following hypothesis:

H_{D12}: Relational satisfaction has a direct positive effect on fairness.

4.3.1.4 The effects of "proactive improvement"

If an LSP displays proactive improvement efforts, this can be expected to have two positive effects. On the one hand, resulting service improvements cause service quality to increase. On the other hand, proactive improvement signals that the LSP will continue to proactively pursue performance enhancements even in the future. While the former, more direct effect, leads to an increased attractiveness of the LSP when comparing B with CL/CL_{alt}, therefore fostering relationship beneficialness (Wallenburg 2004, pp. 216-217), the longer-term signaling effect assures that the LSP will maintain a comparative advantage over alternative LSPs (Wallenburg 2004, pp. 102-103). Accordingly, the following hypotheses can be formulated:

$H_{D13a\text{-}c}$: Proactive improvement has a direct positive effect on (a) repurchases, (b) additional purchases, and (c) referrals.

H_{D14}: Proactive improvement has a direct positive effect on service quality.

In addition, proactive improvement increases perceived fairness. As stated earlier, fairness is achieved when both parties' outcomes, i.e. their input-output-ratios, are balanced. As pointed out above, proactive improvement has a direct, performance increasing effect, which should be perceived by an LSP's customer to render the output component more beneficial. The following hypothesis will therefore be proposed:

H_{D15}: Proactive improvement has a direct positive effect on fairness.

4.3.1.5 The effects of "fairness"

As posited by equity theory, the absence of fairness in a relationship can lead to severe behavioral consequences, such as the termination of a relationship. Accordingly, the fairer the relationship between an LSP and its customer is perceived, the less reason will the customer have to terminate the relationship (Wallenburg 2004, pp. 103-104). Hence, customer loyalty is increased by fairness and the following hypothesis can be formulated:

$H_{D16a\text{-}c}$: Fairness has a direct positive effect on (a) repurchases, (b) additional purchases, and (c) referrals.

Also, commitment is created in fair relationships. As pointed out in chapter 4.2.4, commitment requires the desire to maintain a relationship indefinitely – and this can only be true if parties feel fairly treated. On the contrary, if inequity is perceived, commitment to the relationship decreases. This is reflected in the following hypothesis:

H_{D17}: Fairness has a direct positive effect on commitment.

In addition, fairness will have a positive effect on trust. As trust may be limited by the fear that the other party may not be acting benevolently, continuous fairness in a relationship signals that parties do not intend to behave to the other party's disadvantage. The following hypothesis is therefore formulated:

$H_{D18a\text{-}b}$: Fairness has a direct positive effect on (a) personal trust and (b) organizational trust.

4.3.1.6 The effects of "commitment"

According to the commitment trust theory, commitment is in the center of any buyer-seller relationship. In fact, Morgan and Hunt (1994), as well as other authors (e.g. Too, Souchon, and Thirkell 2001) show that commitment positively influences repurchases, or – as they refer to it – retention. While they do not explicitly examine additional purchases and referrals, Wallenburg (2004, p. 237) validates strong influences also on those two dimensions. The following hypothesis is therefore formulated:

$H_{D19a\text{-}c}$: Commitment has a direct positive effect on (a) repurchases, (b) additional purchases, and (c) referrals.

4.3.1.7 The effects of "trust"

Another central aspect of the commitment trust theory is trust. However, Morgan and Hunt (1994) do not show a direct link between trust and repurchases. Instead, they only propose an indirect effect through commitment as a mediating variable. On the contrary, several authors (e.g. Shemwell, Cronin, and Bullard 1993, Garbarino and Johnson 1999, Chaudhuri and Holbrook 2001, Sirdeshmukh, Singh, and Sabol 2002, Wallenburg 2004) argue for a direct influence of trust on loyalty. Since trust has a risk reducing effect, remaining with a supplier whom one trusts seems to be more logical than switching to an unknown alternative supplier. This study therefore supposes a direct influence of trust on loyalty and the following hypotheses are proposed:

$H_{D20a\text{-}c}$: Personal trust has a direct positive effect on (a) repurchases, (b) additional purchases, and (c) referrals.

$H_{D21a\text{-}c}$: Organizational trust has a direct positive effect on (a) repurchases, (b) additional purchases, and (c) referrals.

Although this study separates the trust construct into organizational and personal trust, they cannot be suspected to be unrelated. Rather, organizational trust will be influenced by personal trust (Wallenburg 2004, p. 220). In long-term relationships between LSPs and their customers, the way in which customers trust their LSPs' organizations is highly influenced by experiences with the individuals at the LSP (see Doney and Cannon 1997). This link is reflected in the following hypothesis:

H_{D22}: Personal trust has a direct positive effect on organizational trust.

A positive influence of trust on commitment is at the core of the commitment trust theory and was validated by Morgan and Hunt (1994). As researchers generally agree about this link (e.g. Wetzels, de Ruyter, and van Birgelen 1998, Sharma and Patterson 1999, Too, Souchon, and Thirkell 2001), the following hypotheses are formulated:

H_{D23}: Personal trust has a direct positive effect on commitment.

H_{D24}: Organizational trust has a direct positive effect on commitment.

4.3.1.8 The effects of "alternatives"

Switching from the current LSP to another LSP is only possible if adequate alternatives exist (Wallenburg 2004, pp. 116-117). Empirically, the availability of alternative LSPs that are expected to offer satisfactory performance was shown to have a negative effect on customer loyalty by

Sriram and Mummalaneni (1990). The following hypothesis is therefore formulated:

H_{D25a-c}: Alternatives have a direct negative effect on (a) repurchases, (b) additional purchases, and (c) referrals.

It was already made clear in chapter 4.2.6 that alternatives are central to social exchange theory as they are used as a comparison level to the existing relationship. Especially the evaluation of satisfaction (in its three dimensions) and proactive improvement is based on comparisons with the offerings of alternative LSPs. For this reason, judgments on satisfaction and proactive improvement will be inferior if adequate alternatives exist. Therefore, alternatives are hypothesized to have a direct negative effect on proactive improvement, service quality, price satisfaction, and relational satisfaction (Wallenburg 2004, pp. 216-218):

H_{D26}: Alternatives have a direct negative effect on proactive improvement.

H_{D27}: Alternatives have a direct negative effect on service quality.

H_{D28}: Alternatives have a direct negative effect on price satisfaction.

H_{D29}: Alternatives have a direct negative effect on relational satisfaction.

4.3.2 Overview of hypotheses and proposed model

Table 4-1 provides an overview of the hypotheses formulated in the preceding section. Graphically, these hypotheses lead to the structural model depicted in Figure 4-3 (effects on customer loyalty are not included for reasons of clarity). This model will be empirically scrutinized in the following chapters.

4.4 Moderating effects of relationship characteristics

Context adequate management has been subject to examination in an extant body of literature that aims at determining factors leading to differences in management practices. While a number of studies have shown that some exogenous factors moderate the constitution of customer loyalty in business relationships (e.g. de Ruyter and Bloemer 1999; Mittal and Kamakura 2001; Homburg, Giering, and Menon 2003), very few authors have yet analyzed how the specific relationship between LSPs and their customers is moderated by relationship characteristics.

Table 4-1. Overview of hypotheses for customer loyalty model

Effects of \ Effects on	Repurchase	Additional Purchase	Referrals	Service Quality	Price Satisfaction	Relational Satisfaction	Proactive Improvement	Fairness	Commitment	Personal Trust	Organizational Trust
Service Quality	H_{D1a}	H_{D1b}	H_{D1c}		H_{D4}	H_{D5}		H_{D6}	H_{D2}	H_{D3a}	H_{D3b}
Price Satisfaction	H_{D7a}	H_{D7b}	H_{D7c}					H_{D8}			
Relational Satisfaction	H_{D9a}	H_{D9b}	H_{D9c}					H_{D12}	H_{D10}	H_{D11a}	H_{D11b}
Proactive Improvement	H_{D13a}	H_{D13b}	H_{D13c}	H_{D14}				H_{D15}			
Fairness	H_{D16a}	H_{D16b}	H_{D16c}						H_{D17}	H_{D18a}	H_{D18b}
Commitment	H_{D19a}	H_{D19b}	H_{D19c}								
Personal Trust	H_{D20a}	H_{D20b}	H_{D20c}						H_{D23}		H_{D22}
Organizational Trust	H_{D21a}	H_{D21b}	H_{D21c}						H_{D24}		
Alternatives	H_{D25a}	H_{D25b}	H_{D25c}	H_{D27}	H_{D28}	H_{D29}	H_{D26}				

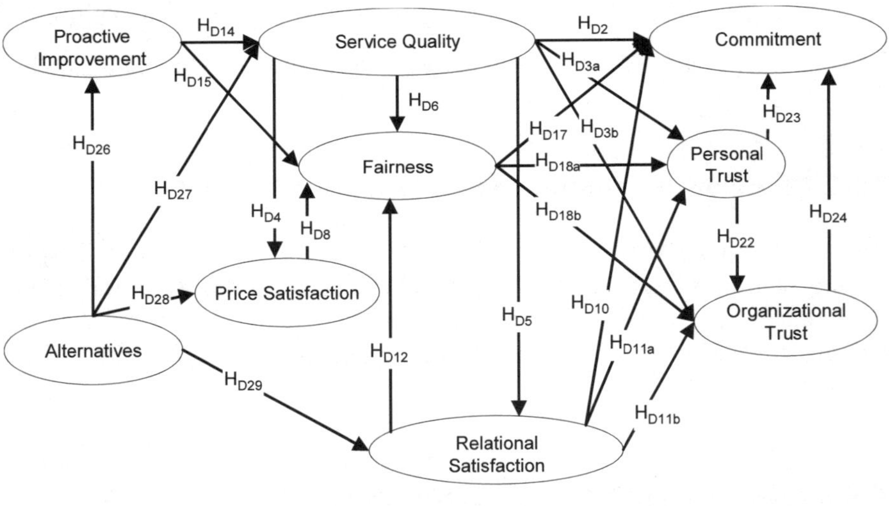

[All determinants are hypothesized to directly affect the three customer loyalty dimensions]

Fig. 4-3. Structural model of determinants of customer loyalty

Taking into account theoretical considerations as well as empirical results obtained by Knemeyer and Murphy (2005), who found some moderating effects of relationship characteristics on the formation of customer loyalty within 3PL arrangements, relationship characteristics can be expected to affect the causal linkages hypothesized in the customer loyalty model presented in the preceding section. In fact, one of the foundational premises of social exchange theory is that exchange interactions over time lead to relational exchange norms governing an exchange relationship (Lambe, Wittmann, and Spekman 2001). Therefore, this section will conceptualize relationship characteristics that are surmised to be of decisive importance for cultivating such exchange norms. These are (1) opportunism, (2) relationship age, (3) centralization of logistics decisions, and (4) outsourcing focus. Then, hypotheses will be formulated regarding the moderating effects of these factors on the links between customer loyalty and its determinants.

4.4.1 Conceptualization of moderators

4.4.1.1 Opportunism

The notion of opportunism stems from transaction cost theory. This theory, which is an important part of the new institutional economics, goes back to Coase (1937), who stated that "the operation of a market costs something" (p. 392), e.g. in terms of the costs of searching out trading partners and negotiating the terms of trade. Later, Williamson concretized the concept of transaction costs, elaborating that informational, contractual, and transactional deficiencies foster opportunistic behavior (e.g. Williamson 1975). Within logistics outsourcing literature, the importance of transaction costs for initial outsourcing decisions, as well as intra-relational governance norms is undisputed (Maltz 1994, p. 260, van Hoek 2000).

An often cited definition of opportunistic behavior is provided by John (1984, p. 279), who maintains that "the essence of opportunistic behavior is the deceit-oriented violation of implicit or explicit promises about one's appropriate or required role behavior". In the context of the present study, opportunism is understood as the reflection of broken promises and false claims by the LSP that weaken the customer's confidence in the future prospects of the relationship.

Several authors, including Morgan and Hunt (1994) and Kwon and Suh (2005), show that opportunistic behavior by a service provider has serious and harmful effects on the trust extended by a customer and ultimately also for customer loyalty. If a customer suspects its LSP to be opportunistic, it has to spend considerable resources on "control and monitoring, resources that could have been deployed more productively for other purposes." (Wathne and Heide 2000, p.36). Also, Calfee and Rubin (1993, p.164) caution that valuable deals may be forgone as a result of opportunism, resulting in substantial opportunity costs. Since opportunism is regarded as "detrimental to logistics outsourcing relationships" by Knemeyer and Murphy (2004, p.41), its moderating effects on the linkages between customer loyalty and its proposed determinants will be examined.

4.4.1.2 Relationship age

Relationship age denotes the period of time that buyers have worked together with an LSP. As maintained by social exchange theory, relationships develop over time and repeated interaction with satisfactory outcomes reinforces bonds between relationship parties and thus increase customer loyalty (Anderson and Sullivan 1993). In fact, Kalwani and Narayandas (1995) point out that decisions on relationship maintenance are not only influenced by recent behavior, but by the experiences made during the entire relationship. With regard to this study, it will be interesting to examine if the formation of customer loyalty differs depending on whether relationship parties have worked together for a long time, or not.

4.4.1.3 Centralization of logistics decisions

Another contingency factor to be analyzed within this study is concerned with the degree of centralization (Ford and Slocum Jr. 1977, p. 562, Child 1972, p. 164), i.e. the locus of logistics outsourcing decisions. These lie between the two poles of fully centralized decisions, where all logistics outsourcing decisions are taken centrally at corporate headquarters, and fully decentralized decisions, where each operating unit takes its own logistics outsourcing decisions.

Centralized structures are characterized by faster decision making processes, better knowledge of the system as a whole, and a higher awareness of market developments. A drawback of centralized decision structures lies in the lower degree of involvement with local/remote situations. Regarding logistics outsourcing, centralized structures lead to more impersonal relationships with an LSP's employees on the working level, and more intense relationships with hierarchically higher levels at the LSP. Consequently, relationships with LSPs can be expected to differ depending on the customer's centralization of logistics decisions.

4.4.1.4 Outsourcing focus

The outsourcing focus refers to the part of the customer's supply chain where the outsourced service is rendered. This can either be on the inbound side, comprising procurement related activities, on the outbound side, with distribution related activities, or internal, i.e. services concerned with the customer's manufacturing or value creating activities. Inbound and outbound focused logistics services are the most often provided services, as shown in Figure 5-2 and Figure 5-4. Usually, they only require a relatively low degree of integration of the LSP into the customer's value creation activities. On the contrary, internal logistics services, comprising e.g. assembly tasks or the provision of supply for the customer's assembly lines, necessitate close integration of LSP and customer and expose the latter to a considerable extent of vulnerability.

While inbound and outbound services should be sensed as important annexes to customers' businesses, firm boundaries are usually not crossed. This is the case, however, in internal arrangements, where customers' own value creating activities usually depend on the internal logistics service rendered by an LSP. In this sense, inbound and outbound oriented arrangements are considerably different from internal arrangements. While the latter lead to vulnerability and eventually to dependence on the specific LSP, an LSP that renders inbound or outbound services is more easily substitutable. For this reason, the formation of customer loyalty can be surmised to be decisively influences by the outsourcing focus, i.e. whether arrangements include inbound and outbound (in the following subsumed under 'external') or internal logistics activities.

4.4.2 Hypotheses for moderating effects

Knemeyer and Murphy (2005) find that several relationship characteristics significantly impact customer loyalty. On the basis of the conceptualizations provided in the preceding paragraphs, this section will propose hypotheses on how the presented relationship characteristics can be expected to moderate the linkages between customer loyalty and its determinants.

4.4.2.1 Moderation by opportunism

Opportunistic behavior by an LSP can be expected to be harmful to any 3PL relationship. In the absence of other governing norms, a customer has no reason to believe that their LSP acts in their best interest and constantly fears to be deceived by its LSP. Consequently, opportunistic behavior should lead to a strong emphasis on current perceptions. Constantly monitoring service quality and evaluating price satisfaction and relational satisfaction thus provides the only safeguard against exploitation, and a relationship will only be maintained and expanded, if the customer remains satisfied. While it may be doubted that a customer would refer an opportunistically behaving LSP at all, satisfaction is hypothesized to at least be a strong prerequisite for recommendation behavior under opportunism, such that:

$H_{M1a\text{-}c}$: The higher the level of opportunism, the stronger is the positive effect of price satisfaction on (a) repurchases, (b) additional purchases, and (c) referrals.

$H_{M2a\text{-}c}$: The higher the level of opportunism, the stronger is the positive effect of service quality on (a) repurchases, (b) additional purchases, and (c) referrals.

$H_{M3a\text{-}c}$: The higher the level of opportunism, the stronger is the positive effect of relational satisfaction on (a) repurchases, (b) additional purchases, and (c) referrals.

As sufficiently elaborated before, trust in an LSP develops out of the customer's conviction that the LSP is both credible and benevolent and has a positive effect on loyalty, because it can limit the perception of risk resulting from uncertainty. However, Morgan and Hunt (1994) as well as Knemeyer and Murphy (2004) have already shown that opportunism is an antecedent to trust, because customers will neither perceive their opportunistic LSPs to be credible nor benevolent. Because of this causal relationship, a moderating effect of opportunism on the trust-loyalty linkage cannot be expected – rather, if included in the model – opportunism should have an effect on loyalty that is mediated through trust. The same should be true for fairness, which, as a reflection of equity perceived in the relationship, should be the lower, the higher the LSP's opportunism is. A moderating effect on the fairness-loyalty link should therefore not be supposed.

Similarly, a causal relationship can be presumed to exist between opportunism and commitment. Even though Morgan and Hunt (1994) only propose a mediated effect of opportunism on commitment, the definition of (affective) commitment used in this study, stating that commitment is the reflection of a customer's desire to maintain a valued relationship, suggests a direct effect. If an LSP is feared to behave opportunistically, less value will be ascribed to the relationship and commitment will thus be lower. Hence, a moderation of the commitment-loyalty link cannot be surmised.

Proactive improvement has a dual character. One result of an LSP's improvement efforts manifests itself in the melioration of service quality. This tangible effect is complemented by a positive signaling effect, when the LSP proves to act proactively in the customer's best interest. While service quality improvements should clearly evoke a positive response from a customer that fears its LSP could be opportunistic, the long-term oriented signaling effect of proactive improvement on loyalty should be less pronounced under opportunism, because a strong focus on current evaluations of satisfaction is assumed. Tending to these ambiguous influences, the following undirected hypothesis is proposed:

$H_{M4a\text{-}c}$: The level of opportunism moderates the effect of proactive improvement on (a) repurchases, (b) additional purchases, and (c) referrals.

Finally, alternatives are hypothesized to have a negative effect on customer loyalty – only if adequate alternatives to the currently used LSP exist, is switching behavior possible. Since opportunism is a harmful trait, diminishing the quality and prospects of a relationship, switching to another LSP can be expected to become a viable option for customers faced with opportunistic LSPs. Therefore, alternatives should have a stronger impact on repurchases, reflecting the increased attractiveness of the option to terminate the relationship. Similarly, alternative LSPs should be more appealing when additional projects are awarded, and the likelihood of referring an opportunistic LSP should decrease. Therefore, the following hypothesis is proposed:

$H_{M5a\text{-}c}$: The higher the level of opportunism, the stronger is the negative effect of alternatives on (a) repurchases, (b) additional purchases, and (c) referrals.

4.4.2.2 Moderation by relationship age

Homburg, Giering, and Menon (2003) already show that relationship age moderates the satisfaction-loyalty link in general. In aged relationships, customers dispose of a wide breadth of experiences with the LSP that must be deemed satisfactory overall – if this were not the case, the relationship would have probably been terminated earlier. Therefore, the way in which dissatisfaction is handled should differ between aged and young relationships. Dissatisfaction in aged relationships is evaluated taking into consideration all positive experiences made in the past. Contrarily, dissatisfaction in young relationships cannot be put into perspective, as customers have to rely on current perceptions of satisfaction only, and dissatisfaction should therefore trigger a stronger response. Hence, the following hypotheses are proposed:

$H_{M6a\text{-}c}$: The higher the age of the relationship, the weaker is the positive effect of price satisfaction on (a) repurchases, (b) additional purchases, and (c) referrals.

$H_{M7a\text{-}c}$: The higher the age of the relationship, the weaker is the positive effect of service quality on (a) repurchases, (b) additional purchases, and (c) referrals.

$H_{M8a\text{-}c}$: The higher the age of the relationship, the weaker is the positive effect of relational satisfaction on (a) repurchases, (b) additional purchases, and (c) referrals.

On the other hand, relationship age is expected to foster trust. As a result of ample interaction over a long period of time that has been – as pointed out above – positive overall, an LSP has had sufficient opportunity to nurture trust through displays of benevolence and credibility. As posited by social exchange theory, trust can serve as a relational exchange norm promoting loyalty towards a service provider. Given that trust needs time to grow, the importance attributed to trust in loyalty considerations should be higher, the longer customers and LSPs have worked together. When relationship age is low on the other hand, time was not sufficient to facilitate the development of trust, such that customers will not be able to rely on trust as a governing norm just as they could if they had cooperated for an extended period of time. This is formally stated for personal and organizational trust as:

$H_{M9a\text{-}c}$: The higher the age of the relationship, the stronger is the positive effect of personal trust on (a) repurchases, (b) additional purchases, and (c) referrals.

$H_{M10a\text{-}c}$: The higher the age of the relationship, the stronger is the positive effect of organizational trust on (a) repurchases, (b) additional purchases, and (c) referrals.

Analogous to the deliberations undertaken for trust, affective commitment and fairness grow with time, as a customer becomes truly convinced that the relationship to its LSP is valuable and worth maintaining, and that outcomes are balanced for both parties. Therefore, commitment and fairness should be more decisive drivers of loyalty in aged than in young relationships, and:

$H_{M11a\text{-}c}$: The higher the age of the relationship, the stronger is the positive effect of commitment on (a) repurchases, (b) additional purchases, and (c) referrals.

$H_{M12a\text{-}c}$: The higher the age of the relationship, the stronger is the positive effect of fairness on (a) repurchases, (b) additional purchases, and (c) referrals.

Due to its dual character, a directed hypothesis cannot be stated for proactive improvement. While customers should esteem proactive improvement's positive signaling effect more highly, the older a relationship is, service quality was argued above to become less important for loyalty with increasing relationship age. The following hypothesis reflects the ambiguous moderation:

$H_{M13a\text{-}c}$: The age of a relationship moderates the effect of proactive improvement on (a) repurchases, (b) additional purchases, and (c) referrals.

Finally, alternatives are surmised to be less important in aged relationships, because a well working, established relationship should be preferable to switching to another provider:

$H_{M14a\text{-}c}$: The higher the age of the relationship, the weaker is the negative effect of alternatives on (a) repurchases, (b) additional purchases, and (c) referrals.

4.4.2.3 Moderation by the centralization of logistics decisions

Outsourcing decisions taken by a centralized department should differ from decisions taken by the operating units directly involved in the respective outsourcing projects. Centralized departments have the disadvantage of not being directly and operationally involved in the respective logistics activities and day-to-day working contact is not common. Therefore, centralized departments should have to rely on aspects that can be monitored easily for their assessment of relationships beneficialness, i.e. satisfaction with price and performance, such that:

$H_{M15a\text{-}c}$: The direct positive effect of price satisfaction on (a) repurchases, (b) additional purchases, and (c) referrals is stronger when the customer's logistics decisions are centralized, than when they are decentralized.

$H_{M16a\text{-}c}$: The direct positive effect of service quality on (a) repurchases, (b) additional purchases, and (c) referrals is stronger when the customer's logistics decisions are centralized, than when they are decentralized.

Similarly, centralized departments cannot rely on personal relationships with an LSP to the same degree that decentralized departments can. As a result, all factors that can only develop from repeated interaction will loose relevance with regard to the formation of customer loyalty. Consequently, commitment and personal trust, as well as relational satisfaction and fairness, should have a weaker influence on customer loyalty, when logistics decisions are centralized, than when they are decentralized, or formally stated as hypotheses:

$H_{M17a\text{-}c}$: The direct positive effect of commitment on (a) repurchases, (b) additional purchases, and (c) referrals is weaker when the customer's logistics decisions are centralized, than when they are decentralized.

$H_{M18a\text{-}c}$: The direct positive effect of personal trust on (a) repurchases, (b) additional purchases, and (c) referrals is weaker when the customer's logistics decisions are centralized, than when they are decentralized.

$H_{M19a\text{-}c}$: The direct positive effect of relational satisfaction on (a) repurchases, (b) additional purchases, and (c) referrals is weaker when the customer's logistics decisions are centralized, than when they are decentralized.

$H_{M20a\text{-}c}$: The direct positive effect of fairness on (a) repurchases, (b) additional purchases, and (c) referrals is weaker when the customer's logistics decisions are centralized, than when they are decentralized.

On the other hand, centralized departments usually dispose of a higher degree of market knowledge, as they are not engaged in daily tasks and possess experience from various outsourcing projects. While personal trust is not significant in this situation, as argued above, organizational trust plays a more important role. Also, much more than decentralized departments will centralized departments therefore assess if alternative LSPs provide better services or lower prices. These considerations lead to the following hypotheses:

$H_{M21a\text{-}c}$: The direct positive effect of organizational trust on (a) repurchases, (b) additional purchases, and (c) referrals is stronger when the customer's logistics decisions are centralized, than when they are decentralized.

$H_{M22a\text{-}c}$: The direct negative effect of alternatives on (a) repurchases, (b) additional purchases, and (c) referrals is stronger when the customer's logistics decisions are centralized, than when they are decentralized.

Finally, centralized departments are surmised to have difficulties in assessing LSPs' proactive improvement efforts. While they should be able to observe resulting performance improvements, they should neither be able to attribute this to proactive improvement efforts, nor should they be able to detect the positive signaling effect. Therefore, proactive improvement should have a weaker influence on loyalty when outsourcing decisions are taken centrally, such that:

$H_{M23a\text{-}c}$: The direct positive effect of proactive improvement on (a) repurchases, (b) additional purchases, and (c) referrals is weaker when the customer's logistics decisions are centralized, than when they are decentralized.

4.4.2.4 Moderation by outsourcing focus

With internal services, LSPs cross the firm boundaries of their customers, getting deeply integrated into the customers' value creation activities, such that their vulnerability and dependence on the LSP are increased. As posited by social exchange theory, customers should associate vulnerability and dependence with risk, which can be mitigated by trust. Therefore, trust should have an increased importance for loyalty in internally focused outsourcing projects in contrast to externally focused project, and:

$H_{M24a\text{-}c}$: The direct positive effect of personal trust on (a) repurchases, (b) additional purchases, and (c) referrals is stronger when the outsourcing focus is internal than when it is external.

$H_{M25a\text{-}c}$: The direct positive effect of organizational trust on (a) repurchases, (b) additional purchases, and (c) referrals is stronger when the outsourcing focus is internal than when it is external.

Being deeply implanted in their customers' value creation activities, LSPs are harder to replace in internally focused projects than in externally focused projects. Consequently, switching is aggravated and alternatives should be less relevant for customers in internal projects, such that the following hypothesis can be proposed:

$H_{M26a\text{-}c}$: The direct negative effect of alternatives on (a) repurchases, (b) additional purchases, and (c) referrals is weaker when the outsourcing focus is internal than when it is external.

As pointed out above, customers that allow LSPs to cross their firm boundaries by outsourcing part of their value creation activities are exposed to risk, resulting from vulnerability and dependence. As postulated by social exchange theory, customers can only be expected to do so, if relational governance structures have been established prior to outsourcing internal activities. A prerequisite of the development of such governance structures is the existence of a history of cooperation with satisfactory and beneficial outcomes. For this reason, short-term dissatisfaction with certain issues of the cooperation should not have serious detrimental effects, as they would be put into perspective by past satisfactory interaction. Consequently, the perception of satisfaction can be surmised to have a weaker effect in such internal projects, in contrast to external projects, where the relationship may not be equally well established. This is reflected by the following hypotheses:

$H_{M27a\text{-}c}$: The direct positive effect of price satisfaction on (a) repurchases, (b) additional purchases, and (c) referrals is weaker when the outsourcing focus is internal than when it is external.

$H_{M28a\text{-}c}$: The direct positive effect of service quality on (a) repurchases, (b) additional purchases, and (c) referrals is weaker when the outsourcing focus is internal than when it is external.

$H_{M29a\text{-}c}$: The direct positive effect of relational satisfaction on (a) repurchases, (b) additional purchases, and (c) referrals is weaker when the outsourcing focus is internal than when it is external.

Fairness, on the other hand, is a result of repeated satisfactory assessments of a relationship's outcomes, i.e. the input-output-ratio achieved in a relationship. Fairness thus develops as a result of a history of cooperation between LSPs and their customers and can serve as a means to mitigate perceived risk just as commitment and trust do. In an internally focused outsourcing relationships, perceived fairness can therefore be assumed to have a stronger impact on loyalty considerations, than in externally focused arrangements. This argumentation is formally stated in the following hypothesis:

$H_{M30a\text{-}c}$: The direct positive effect of fairness on (a) repurchases, (b) additional purchases, and (c) referrals is stronger when the outsourcing focus is internal than when it is external.

The moderation of the outsourcing focus on the commitment to loyalty link is ambiguous. While on the one hand, commitment should be more important in internal projects relative to external projects, it is not clear, whether this really results from affection towards the LSP. Rather, normative factors can be surmised to prevail as a result of the dependence created in consequence to the portrayed deep integration of the LSP into the customer's value creation activities. A moderation of the link between affective commitment and loyalty can therefore not be ascertained and no hypothesis is stated.

Similarly, no moderation hypothesis is proposed for the effect of proactive improvement on customer loyalty. As shown before, proactive improvement's influence on loyalty is composed of two distinct effects. On the one hand, there is a long-term positive signaling effect, which fosters trust, commitment, and fairness. As argued in hypotheses $H_{M24a\text{-}c}$, $H_{M25a\text{-}c}$, and $H_{M30a\text{-}c}$, this long-term effect should be stronger in internally focused than in externally focused outsourcing arrangements. On the other hand, however, proactive improvement has a short-term performance increasing effect, which should be weaker in internally focused arrangements, as stated in $H_{M28a\text{-}c}$. The argumentation for the moderating effect of the outsourcing focus on the link between proactive improvement and loyalty is therefore ambiguous and consequently, no hypothesis is formulated.

4.4.3 Overview of moderation hypotheses

Table 4-2. Overview of moderation hypotheses

Customer loyalty determinant	Effect of determinant on customer loyalty moderated by...							
	Opportunism		Relationship age		Centralization		Outsourcing focus	
	High vs. Low		High vs. low		High vs. low		In- vs. external	
Service quality	Stronger	$H_{M2a\text{-}c}$	Weaker	$H_{M7a\text{-}c}$	Stronger	$H_{M16a\text{-}c}$	Weaker	$H_{M28a\text{-}c}$
Price satisfaction	Stronger	$H_{M1a\text{-}c}$	Weaker	$H_{M6a\text{-}c}$	Stronger	$H_{M15a\text{-}c}$	Weaker	$H_{M27a\text{-}c}$
Relational satisfaction	Stronger	$H_{M3a\text{-}c}$	Weaker	$H_{M8a\text{-}c}$	Weaker	$H_{M19a\text{-}c}$	Weaker	$H_{M29a\text{-}c}$
Fairness	---		Stronger	$H_{M12a\text{-}c}$	Weaker	$H_{M20a\text{-}c}$	Stronger	$H_{M30a\text{-}c}$
Commitment	---		Stronger	$H_{M11a\text{-}c}$	Weaker	$H_{M17a\text{-}c}$	---	
Personal trust	---		Stronger	$H_{M9a\text{-}c}$	Weaker	$H_{M18a\text{-}c}$	Stronger	$H_{M24a\text{-}c}$
Organizational trust	---		Stronger	$H_{M10a\text{-}c}$	Stronger	$H_{M21a\text{-}c}$	Stronger	$H_{M25a\text{-}c}$
Alternatives	Stronger	$H_{M5a\text{-}c}$	Weaker	$H_{M14a\text{-}c}$	Stronger	$H_{M22a\text{-}c}$	Weaker	$H_{M26a\text{-}c}$
Proactive Improvement	Undirected	$H_{M4a\text{-}c}$	Undirected	$H_{M13a\text{-}c}$	Weaker	$H_{M23a\text{-}c}$	---	

Table 4-2 presents an overview of the moderation hypotheses proposed in the previous sections. While moderating effects are hypothesized for most linkages between customer loyalty and its determinants, it was not possible to provide stringent argumentation for the influence of the respective moderators on six out of 36 loyalty-effects. It was especially difficult to determine moderating influences on the effect of proactive improvement on customer loyalty, due to proactive improvement's dual character. This is reflected in the two undirected hypotheses ($H_{M4a\text{-}c}$, $H_{M13a\text{-}c}$), where it was impossible to determine, which of the two consequences of proactive improvement would be moderated by the two respective moderators. For the remaining 28 relationships, the direction of the moderation was identified and consequently stated in the hypotheses.

4.5 Moderating effects of national culture

It is the aim of the present section to point out differences between Germany and the USA regarding the formation of customer loyalty as a result of national culture. To achieve this, cultural differences between Germany and the USA, as conceptualized in chapter 3.4, will be applied to derive a set of hypotheses regarding the linkages between customer loyalty and its determinants.

4.5.1 Hypotheses on the basis of cultural differences

As shown in Table 3-1, ten cultural differences between the USA and Germany were extracted from the frameworks of Hofstede (2001), Hall and Hall (1990), and/or Trompenaars and Hampden-Turner (1997). As all determinants of customer loyalty (service quality, price satisfaction, relational satisfaction, proactive improvement, fairness, commitment, personal and organizational trust, and alternatives) were previously argued to have direct effects on all three dimensions of customer loyalty, these causal linkages will be scrutinized in the following, employing both reasoning on the basis of the identified cultural differences, and applicable research, if that exists.

Both the USA and Germany were shown to be results oriented and therefore prone to maintain a relationship with a service provider if adequate performance is delivered. In fact, masculine cultures such as Germany and the USA, as pointed out by Hofstede (2001, p. 319), see low performance, i.e. insufficient service quality, as a main reason to terminate a relationship.

Several empirical studies have already tackled the role of service quality in intercultural settings and come to the conclusion that the perception of service quality varies between different cultures (e.g. Mattila 1999, Winsted 1999, Sultan and Simpson Jr. 2000, Ueltschy and Krampf 2001, Witkowski and Wolfinbarger 2002). Liu, Furrer, and Sudharshan (2001) and Laroche et al. (2004) even consider the effect of service quality on customer loyalty with regard to cultural differences and Furrer, Liu, and Sudharshan (2000, p. 355) actually discover a positive correlation between Hofstede's cultural dimensions and service quality. They conclude that highly individualistic cultures, like the USA, "due to their drive and self-responsibility ethic, demand that others be efficient and therefore demand a high level of service quality." (Furrer, Liu, and Sudharshan 2000, p. 363) In their later study, Liu, Furrer, and Sudharshan (2001, p. 126) empirically validate that individualists like the US-Americans "exhibit a higher intention to switch to another service provider or to give negative word of mouth if they experience negative service quality" and that uncertainty avoiding cultures like the Germans "show a lower intention to switch" and "to give negative word of mouth" if negative service quality is experienced. This is further supported by Sultan and Simpson Jr. (2000) and Witkowski and Wolfinbarger (2002), who find that service expectations are generally lower in Germany than in the USA. Consequently, the following hypothesis can be formulated:

$H_{C1a\text{-}c}$: The direct positive effect of service quality on (a) repurchases, (b) additional purchases, and (c) referrals is stronger in the USA than in Germany.

The authors of the three cultural frameworks considered in chapter 3.4 agree that the USA are a lot more focused on achieving immediate economic benefits whenever they engage in business relationships. It is actually accepted in the USA to terminate a relationship, if an alternative setting promises to yield a better return. On the contrary, Germans value stable and long-term relationships and feel a strong sense of loyalty and responsibility, if the other party in the relationship has been admitted to the in-group. While price satisfaction is assumed to play an important role in both countries, its effect on customer loyalty should be even higher in the USA than in Germany, leading to the following hypothesis:

$H_{C2a\text{-}c}$: The direct positive effect of price satisfaction on (a) repurchases, (b) additional purchases, and (c) referrals is stronger in the USA than in Germany.

Americans genuinely believe that conflict and competition are constructive, necessary, and useful, while Germans appreciate consensus and dread open conflict, as they are seen as potential sources of ambiguity. In addition, Germans more readily share information and foster mutually beneficial cooperation, which can be assumed to increase satisfaction with the relationship (Sattler, Schrader, and Lüthje 2003, pp. 279-280). Also, personal bonds as re-enforcers of business relationships are more important in Germany than in the USA, such that relational satisfaction can be expected to be a stronger driver of loyalty in Germany. The following hypothesis can therefore be proposed:

$H_{C3a\text{-}c}$: The direct positive effect of relational satisfaction on (a) repurchases, (b) additional purchases, and (c) referrals is weaker in the USA than in Germany.

Proactive improvement, as stated above, affects customer loyalty, because it has two distinct effects. In the short term, proactive improvement meliorates the performance provided by the LSP. In the long term, proactive improvement signals that continuous efforts are invested into the relationship, which will make the LSP remain beneficial compared to alternatives. While the more short-term performance-increasing effect appeals to both Germans and Americans, because they both value immediate results, the long-term effect evokes a much stronger reaction in Germany. As Germans are long-term oriented, preferring long-lasting, stable relationships, they will appreciate proactive improvement more than Americans will.

On the other hand, Germany is a rather risk avoiding country, in which change and innovation are dreaded. In addition, Germans strictly adhere to rules and regulations, such as those found in contractual agreements between LSPs and their customers. For this reason, Germans may actually not see the benefits of proactive improvement, as change is effectuated and a deviation from initially agreed upon rules is induced. Overall, the effect culture should have on the causal linkage between proactive improvement and loyalty is ambiguous and a difference can therefore not be hypothesized.

The examination of German and American culture in chapter 3.4 revealed that both Germany and the USA universally require equal treatment, cherish fairness and object inequalities. For this reason, fairness can only be expected to be equally important in the USA and Germany. This argument is also supported by previous research, which discovered that "people from different cultures generally agree that justice is important" and that "interest in justice seems to be universal." (Greenberg 2001, pp. 370-371) In addition, Austen (2002), in a study on the acceptance of wage differences between high- and low-status workers, found that Germany and the USA are basically similar in their tolerance for distributive unfairness.

Consequently, it is not reasonable to assume differences between Germany and the USA regarding the causal linkage between fairness and customer loyalty and the following hypothesis is postulated:

$H_{C4a\text{-}c}$: The direct positive effect of fairness on (a) repurchases, (b) additional purchases, and (c) referrals is equal in the USA and Germany.

As outlined before, Germans are more uncertainty avoiding than US-Americans and strive to reduce uncertainty and ambiguity by engaging in long-term relationships[2]. It was also shown in chapter 3.4 that Germany is more collectivistic than the USA and that Germans feel mutual responsibility towards in-group members. Since long-term relationships in Germany are oftentimes facilitated by personal bonds and accepting the respective other party of the relationship to one's own in-group is common, business relationships in Germany usually include an emotional component with mutual affective commitment. Contrariwise, the US society is marked by short-term thinking and a focus on immediate monetary payoffs. Also, personal bonding is uncommon in the highly specific US culture and accordingly, business relationships in the USA usually lack such an affective component. For this reason, affective commitment can be expected to be very important in Germany, while it should be of subordinate value for explaining customer loyalty in the USA. This is stated in the following hypothesis:

$H_{C5a\text{-}c}$: The direct positive effect of affective commitment on (a) repurchases, (b) additional purchases, and (c) referrals is weaker in the USA than in Germany.

Based on the analysis of US and German cultures, differences in the effect of trust on customer loyalty can be expected. As stated above, business relationships in Germany are often accompanied by personal relationships and mutual admittance of relationship partners to the respective in-groups. According to Hofstede, Germans extend trust towards in-group members more easily than Americans, while Germans may be hesitant about trusting out-group members. Americans, on the other hand, are impersonal in the business world, rather calculative, and hence less likely to trust. Another important aspect in Germany's uncertainty avoiding culture is the potential risk minimization reached through maintaining a trustful relationship. For these reasons, trust should be more important for building customer loyalty in Germany than in the USA.

However, personal and organizational trust have to be distinguished in the cultural context. While Germans extend trust towards in-group members, organizations can never be considered in-group. A prerequisite for admittance to the in-group is the existence of personal bonds and the latter can never be formed with an organization, but exclusively between individuals. Americans, as individualists, do not differentiate between in- and out-groups, such that a distinction between personal and organizational trust cannot be made for the USA. For this reason, personal trust should be

[2] In fact, the empirical study which will be introduced in chapter 5 shows that the average relationship age in Germany is over 7 years, while it is 5.5 years in the USA.

more important in Germany, while organizational trust cannot be assumed to be different between the USA and Germany. The following hypotheses are therefore proposed:

$H_{C6a\text{-}c}$: The direct positive effect of personal trust on (a) repurchases, (b) additional purchases, and (c) referrals is weaker in the USA than in Germany.

$H_{C7a\text{-}c}$: The direct positive effect of organizational trust on (a) repurchases, (b) additional purchases, and (c) referrals is equal in the USA and Germany.

The existence of alternative LSPs that provide services at the same price and quality is a prerequisite for switching behavior. Cultural analysis leaves no doubt that German customers are a lot less likely to turn to an alternative LSP than American customers, because Germans favor stable and long-term relationships, dread change, and try to avoid risk (Money 2004, p. 300). Americans, on the other hand, are willing to take unknown risks if higher pay-offs are expected. In fact, Liu, Furrer, and Sudharshan (2001, p. 126) show empirically that members of highly individualistic societies, such as the USA, have a greater propensity to switch and that uncertainty avoidance prevents switching. Therefore, the following hypothesis is postulated:

$H_{C8a\text{-}c}$: The direct negative effect of alternatives on (a) repurchases, (b) additional purchases, and (c) referrals is stronger in the USA than in Germany.

4.5.2 Overview of hypotheses for the effects of national culture

The preceding section showed that differences between Germany and the USA can be expected with respect to the effects on customer loyalty of six out of nine factors. While service quality, price satisfaction, and alternatives should be more important in the USA, relational satisfaction, commitment, and personal trust should be more important in Germany. Fairness and organizational trust cannot be assumed to have different effects in Germany and the USA and ambiguous evidence was presented for proactive improvement, such that a hypothesis regarding this factor could not be derived. Table 4-3 gives an overview of the proposed hypotheses.

Table 4-3. Overview of hypotheses on cultural differences

Hypothesis	Customer loyalty determinant	Effect on customer loyalty in the USA compared to Germany
$H_{C1a\text{-}c}$	Service quality	Stronger
$H_{C2a\text{-}c}$	Price satisfaction	Stronger
$H_{C3a\text{-}c}$	Relational satisfaction	Weaker
$H_{C4a\text{-}c}$	Fairness	Equal
$H_{C5a\text{-}c}$	Commitment	Weaker
$H_{C6a\text{-}c}$	Personal trust	Weaker
$H_{C7a\text{-}c}$	Organizational trust	Equal
$H_{C8a\text{-}c}$	Alternatives	Stronger

5 Methodology and sample characteristics

This chapter will provide the methodological basis for empirically examining the hypotheses suggested in the preceding chapter. For this, information on survey design, as well as on the samples obtained in Germany and the USA will be provided first. Then, sections 5.3 to 5.5 will give an introduction to the data analysis methodology employed for assessing the hypothesized models, before the chapter will be concluded with deliberations on comparative analyses, as needed for answering research questions three and four.

5.1 Survey design

The present section will provide information on the design of data collection in this study. For that, the research object, the appropriate data collection procedure, and targeted informants will be identified, before details on questionnaire design and data collection will be provided.

5.1.1 Research object

In order to provide answers to the research questions put forward in chapter 2, relationships between suppliers and buyers of logistics services in Germany and the USA are at the core of this research. Since business units within large companies often differ significantly in their use of LSPs, only those business units that actually contract logistics services from LSPs are relevant.

5.1.2 Methods for data analysis

In order to select the appropriate data gathering approach, the goals of the research as well as the employed methods for data analysis have to be considered. Aiming at the examination of relationships between different factors, several methods could be used, of which the most important ones will be outlined in the following. However, these methods are only applicable in the context of the research at hand, if they fulfill the following criteria (Peter 1997, pp. 128-130 and Wallenburg 2004, p. 124):

The hypotheses formulated in chapter 4.3 for the comprehensive model of customer loyalty establish relationships between factors that cannot be measured directly, which are therefore called theoretical constructs (see also chapter 5.3). The first criterion an applicable method has to satisfy therefore is the possibility to incorporate relationships between several constructs.

In order to analyze these theoretical constructs (i.e. latent variables), they have to be measured through empirically observable indicators reflecting the latent variables. However, measurement errors always occur when measuring latent variables and Homburg (1989, p. 20) cautions to carefully evaluate them to avoid drawing false conclusions. It is therefore the second criterion that an applicable method has to take into account measurement errors.

In addition, the proposed model of customer loyalty includes several interdependencies between the potential determinants of customer loyalty (exogenous variables) as well as effects of these on the three dimensions of customer loyalty. For this reason, an applicable method has to be able to evaluate interdependencies between exogenous variables (criterion three). Furthermore, the assessment of the model's goodness-of-fit requires the simultaneous calculation of all hypothesized relationships of the model (criterion four).

In general, the analysis of cause and effect relationships can be conducted by employing multivariate methods that are based on regression analysis (Backhaus, Baumeister, and Mühlfeld 2003, Gujarati 2003, pp. 15-333, and Wallenburg 2004, pp. 124-125). Traditional regression analysis, however, does not meet the required criteria. First, regression analysis can only evaluate observable variables. For this reason, all indicators pertaining to theoretical constructs would have to be aggregated before employing regression analysis (e.g. through factor analysis), leading to the omission of measurement errors. In addition, interdependencies between exogenous variables cannot be assessed, because regression analysis requires the independence of regressors and does therefore not allow the simultaneous evaluation of all hypotheses. Hence, regression analysis cannot be used for the present research.

The logit approach is another potentially appropriate method, because it allows the simultaneous estimation of several hypotheses in a model. However, it is not suitable for the present research, because it does not allow exogenous variables to be interdependent.

According to Wallenburg and Weber (2005, pp. 172-173), the standard procedure for the analysis of cause and effect relationships between theoretical constructs is structural equation modeling (SEM). By combining elements from regression analysis and factor analysis, SEM explicitly considers measurement errors, assesses relationships between exogenous variables, and simultaneously evaluates all hypotheses. In this way, SEM satisfies all required criteria and is appropriate for analyzing the comprehensive model of customer loyalty. In addition, SEM also provides the possibility to carry out comparative analyses and is therefore also suitable for analyzing the moderating effects of contingency factors and cultural context.

5.1.3 Data collection procedure

It is the aim of the present research to draw universal conclusions, such that "research findings can be generalized to the broader population." (Mentzer and Flint 1997, p. 211) This generalization requires high external validity, which can be achieved through a large sample size. The most adequate and, according to Lambert and Harrington (1990, p. 5), predominantly used survey method for obtaining large sample sizes in industrial survey settings is the written survey. Written surveys have several advantages over other survey methods, e.g. interviews or case studies. Written surveys are relatively easy to conduct, comparatively inexpensive, and yield high numbers of cases. Also, respondents are not biased by an interviewer, as it would be the case in interviews or case studies and they can take as much time as they want to to answer the survey. In fact, Mentzer and Kahn (1995, p. 242) find that over 50% of articles published in the Journal of Business Logistics between 1978 and 1993 employ written surveys, while interviews were only used in 13.8% and case studies in 3.2% of cases. Major deficits of written surveys are declining response rates (Baruch 1999), questionnaire length (Yammarino, Skinner, and Childers 1991), and the lack of control (Mentzer and Kahn 1995). Still, written surveys are the most viable form for achieving the research goals of the present study – neither interviews nor case studies would yield the number of cases required to ascertain high external validity.

Having decided for the written survey form, there are two possible ways to conduct the survey – a mail survey or a web-based survey. Griffis, Goldsby, and Cooper (2003) compared these two survey procedures and conclude that web-based surveys, in which sample members are contacted via e-mail and complete the written survey over the internet, are clearly preferable when large samples are contacted. They also show that response rates are slightly higher in web-based surveys, costs are lower for large sample sizes, and quality is comparable to mail surveys (an overview of advantages and disadvantages is also given by Grant, Teller, and Teller 2005, p. 152). As pointed out above, the present research strives to realize a large sample-size, wherefore a web-based survey procedure is chosen.

5.1.4 Informants

Zimmer (2000, p. 130) states that centralized and decentralized survey configurations are feasible for conducting research on business relationships:

Centralized surveys focus on one company and its relationships to other companies. For the research at hand, a single customer could be surveyed about its relationships to different LSPs, or a single LSP could be surveyed about its relationships to different customers. This approach, however, is not suitable for the present study. If only one customer were asked about its relationships, only a very small sample size could be achieved. If only one LSP were asked, generalizations would hardly be possible (Wallenburg 2003, p. 83).

Decentralized surveys include relationships between a multitude of customers and LSPs. In this case, either customers or LSPs could be surveyed regarding their relationships to the respective other parties. For the present research, the decentralized configuration is most suitable, because a large sample size can be achieved and results can be generalized.

In addition to the above outlined uni-lateral survey configurations, dyadic surveys can be conducted, surveying both relationship parties, i.e. LSPs and their customers, with regard to the relationships they share. The current research, however, is focused on the buyers of logistics services, because only their intentions and evaluations are relevant for determining customer loyalty and its determinants (Wallenburg 2004, p. 127). While a dyadic survey may provide some additional insights, it would also complicate data collection and jeopardize sample size. On this account, a unilateral decentralized survey configuration is selected, in which customers are surveyed regarding their relationships to LSPs. In addition, customers are asked to recur only to their most important relationship to an LSP. This focus has the advantage of restricting questionnaire length to a tolerable level and of ensuring the necessary strategic relevance of the issue to the respondents.

Next, relevant informants in customers' organizations have to be identified. Theoretically, everybody dealing with LSPs would be eligible for questioning. Kumar, Stern, and Anderson (1993, p. 1636), however, admit that "how to combine the discrepant responses of multiple informants into an organizational response is an unresolved issue." For this reason, the managers responsible for logistics on the business unit level are selected as key informants. While this single informant approach is sometimes criticized for disregarding potential differences between answers from informants from one company (see e.g. Ernst 2001, p. 7), other authors, such as John and Reve (1982) and Deshpandé, Farley, and Webster Jr. (1993), acknowledge general correspondence between multiple informants' answers. In the context of the present research, logistics managers can be assumed to possess the information necessary to evaluate the relationship with their most important LSP and are eventually relevant with respect to customer loyalty. Even if the manager is not perfectly informed about every detail of the relationship, it is his subjective assessment that facilitates customer loyalty. Despite the criticism, logistics managers therefore are adequate key informants for the present study.

5.1.5 Questionnaire design

The development of the questionnaire was based on the conceptualization of customer loyalty, its determinants, and relationship characteristics, as described in chapter 4. In order to decrease subjectivity and ensure interpersonal comparability, open questions were avoided. Instead, questions were devised as statements and participants of the survey were asked to indicate their level of agreement with each statement on a seven-point Likert-scale[1].

Before commencing the questionnaire, participants were explicitly instructed to select their most important LSP and to answer all questions keeping this LSP in mind. The actual survey then began with general questions about the respondents' firms' logistics and outsourcing activities, before continuing with questions about the relationship to the selected LSP. A third part was concerned with questions about the respondents' situation versus its own customers, competitors, and corporate success. The questionnaire closed with structural questions about the participant and its business unit.

5.1.6 Pretest interviews

While the questionnaire was first developed in German, it was subsequently translated to English for use in the USA. Back-translation to Ger-

[1] A seven-point Likert-scale is a balanced rating scale with seven possible answer values, of which the middle one is neutral.

man was performed "to ensure that the meaning of each question was as consistent as possible across nations." (Huff and Kelley 2005, p. 99)

Following questionnaire design, pretest interviews took place to identify potential shortcomings of the survey instrument. On this account, 10 interviews with professional experts, four interviews with logistics professors, and another eight interviews with other logistics researchers were conducted, both in Germany and the USA. As a consequence, minor modifications were made regarding the wording of the questions that lead to an increased comprehensibility and unambiguousness. All changes were reflected in the German and the English instrument. Overall, interviewees perceived the questionnaire to be well structured and of tolerable length.

5.1.7 Data collection

Most logistics managers in Germany and the USA are organized in the respective professional organizations, i.e. Bundesvereinigung Logistik e.V. (BVL) in Germany, and the Council of Supply Chain Management Professionals (CSCMP) in the USA. For this reason, both organizations were contacted and asked for support regarding the present study. While BVL agreed to provide a sample free of charge, a database was obtained from CSCMP at a small fee. In both cases, data selection could only be influenced with regard to industry affiliation. Consequently, address-data from all manufacturing and trade industries was selected, while industries such as information technology, LSPs, banking, consulting, and insurances were excluded. Names and e-mail-addresses of 4,570 individuals were obtained from BVL, and 1,538 from CSCMP. In addition, the database from the Kuehne-Center of Logistics Management (KLM) could be used in Germany, yielding another 678 individuals, such that the German data set comprised a total of 5,248 names.

Members of the German-sample were sent an initial e-mail signed by Prof. Dr. Jürgen Weber from KLM at the WHU – Otto Beisheim School of Management, which included a hyperlink to the web-survey. The US sample received a similar email signed by Prof. Dr. Juergen Weber and Professors Thomas J. Goldsby and A. Michael Knemeyer of The Ohio State University[2]. Data collection was conducted in Germany first, taking place

[2] Thomas J. Goldsby (now at the Gatton College of Business and Economics, University of Kentucky) and A. Michael Knemeyer of the Department of Marketing and Logistics, Fisher College of Business, The Ohio State University, supported data gathering in the USA.

between October 2004 and January 2005, and afterwards in the USA, between April and June 2005.

In order to increase response rates, two approaches suggested by Larson and Poist (2004) were incorporated. First, incentives were offered to all participating individuals. In Germany, participants had the choice between free participation at a logistics function hosted by KLM and a free book. In the USA, participants had to select between participation in a lottery for an Apple iPod®, participation in a lottery for a faculty hosted on-site seminar, a free book, and a coupon for a 50% discount on a book.

On top, all participants received a personalized management summary of the study's results. The second step for increasing response rate was sending reminder e-mails to everyone that had not participated so far. While the KLM- and the CSCMP-sample received two reminder e-mails, individuals in the BVL-sample only received one[3].

5.2 Samples

After having outlined the design of data collection in the preceding section, the subsequent paragraphs will first characterize the two samples in Germany and the USA, before representativeness of the samples will be discussed.

5.2.1 Sample characteristics in Germany

Out of the 5,248 e-mails sent to the German sample, 1,846 were returned as undeliverable and therefore had to be subtracted from the total sample, resulting in 3,402 e-mails that supposably reached their addressees. From these, a total of 579 questionnaires were completed, resulting in a response rate in Germany of 17.02%. Since SEM does not permit any incomplete data, these 579 responses had to undergo some initial treatment, for which Byrne (2001, pp. 289-291) suggests three procedures: listwise deletion, in which all cases having missing values are excluded from all further analyses (see e.g. Boomsma 1985); pairwise deletion, in which only cases with a missing value for a specific analysis are excluded (see e.g. Marsh 1998); and imputation, i.e. replacing missing values with some estimate. Unfortunately, 34 responses had such a large number of missing values that they had to be eliminated using listwise deletion. For the remaining 545 cases,

[3] BVL only agreed to support the German study under the condition that merely a total of two e-mails were sent to their sample.

pattern-matching imputation was employed, by replacing missing values "with an observed score from another case in the data for which the response pattern across all variables is similar." (Byrne 2001, p. 291) All subsequent analyses are therefore based on this sample of 545 cases.

Table 5-1. Size and industry affiliation of respondents (Germany)

Firm size	Share
< 100 mil. €	20.8%
100-249 mil. €	16.0%
250-499 mil. €	20.0%
500-999 mil. €	14.7%
≥ 1,000 mil. €	28.6%

Industry	Share
Healthcare	5.6%
Food and beverage	6.7%
Consumer goods	8.5%
Industrial equipment	8.5%
Chemicals and plastics	9.8%
Electronics, precision mechanics, optics	14.6%
Retailing	14.6%
Automotive	15.2%
Others	16.5%

The sample contains a wide variety of different types of firms[4] with regard to firm-sizes and industry affiliation. Small and medium sized companies with yearly revenues of less than € 100 million are relatively underrepresented with a share of only 21%, while large companies (at least € 500 million) constitute nearly half of the sample. Looking at industry affiliation, the sample is well balanced and all major industries are represented to a significant extent, as Table 5-1 shows.

Within companies, logistics outsourcing plays a very important role and on average about 45% of total logistics cost is spent on LSPs, with little variance over company sizes. Thereof, the most important LSP, which was selected by respondents as a reference for all relationship specific questions, has a share of ca. 46%. In smaller companies (up to € 250 million in

[4] Although the provided answers refer to business units, the terms "firm" and "company" are used synonymously for reasons of simplicity.

revenues), its share is even 52% and thus higher than in very large companies, where it may fall below 40%. The results are shown in Figure 5-1.

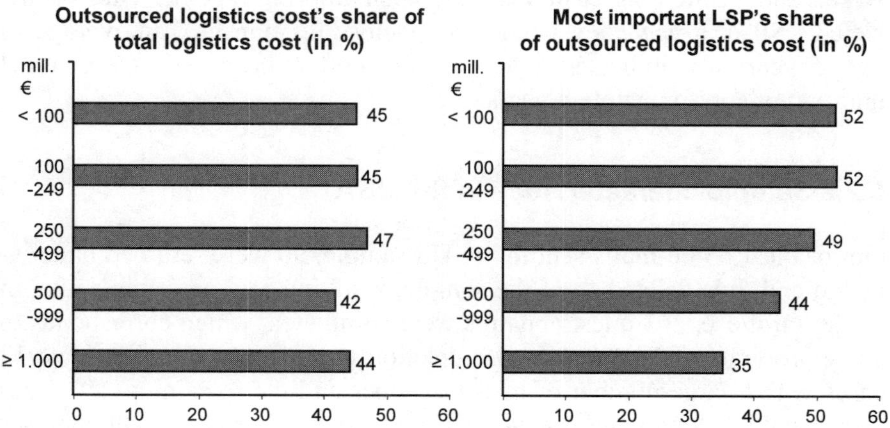

Fig. 5-1. Importance of logistics outsourcing (Germany)

The logistics activities carried out by the selected LSP are mainly restricted to outbound flows. While inbound flows play a minor role, internal logistics activities are rarely outsourced. Of all activities provided by the selected LSP, transportation services are most commonly outsourced, followed by warehousing, and value added services such as inventory management and pick/pack operations. Advanced services, e.g. lead logistics management or assembly services, are seldom outsourced, as shown in Figure 5-2.

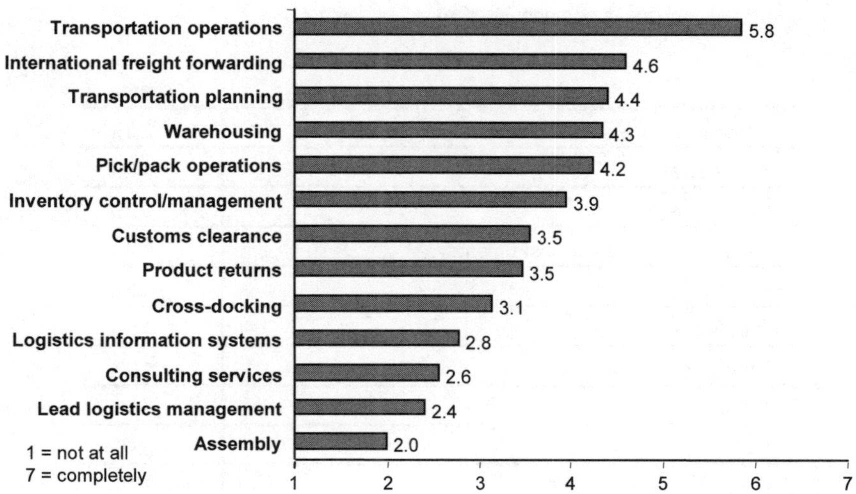

Fig. 5-2. Logistics services provided by the selected LSP (Germany)

Relationships between customers and their most important LSPs are mostly long-term, with an average relationship duration of over 7 years. Asked about how long customers had been working together with the selected LSP in a way they would call a close relationship, an average of over 5 years was indicated, which is still more than the average contract duration of approximately 4 years.

5.2.2 Sample characteristics in the USA

Out of the 1,538 e-mails sent to the US sample, 90 were returned undeliverable and thus deleted from the sample, resulting in a net sample size of 1,448. Of those, 263 questionnaires were completed, which corresponds to a response rate of 18.2%. Listwise deletion was employed to eliminate 13 cases with a large number of missing values and the remaining 250 cases were handled using pattern-matching imputation. The subsequent analyses of the US sample are based on these 250 cases.

Table 5-2. Size and industry affiliation of respondents (USA)

Firm size	Share
< 100 mil. $	6.5%
100-249 mil. $	10.6%
250-499 mil. $	11.8%
500-999 mil. $	10.2%
1,000-4,999 mil. $	29.4%
≥ 5,000 mil. $	31.4%

Industry	Share
Computer hardware and peripheral equipment	2.8%
Automotive	4.8%
Industrial equipment	5.2%
Electronics and related instruments	8.0%
Healthcare	8.8%
Retailing	9.6%
Chemicals and plastics	10.8%
Other	15.3%
Consumer goods	16.9%
Food and beverage	17.7%

With respect to industry representation, the US sample is very well balanced. Regarding firm size, the sample shows a strong focus on very large companies with yearly revenues of at least 1 billion US-$ and a share of over 50%, while small and medium sized companies with revenues of less than 100 million US-$ are underrepresented at a share of only 7%. Industry affiliation and firm size by yearly revenues are graphically shown in Table 5-2.

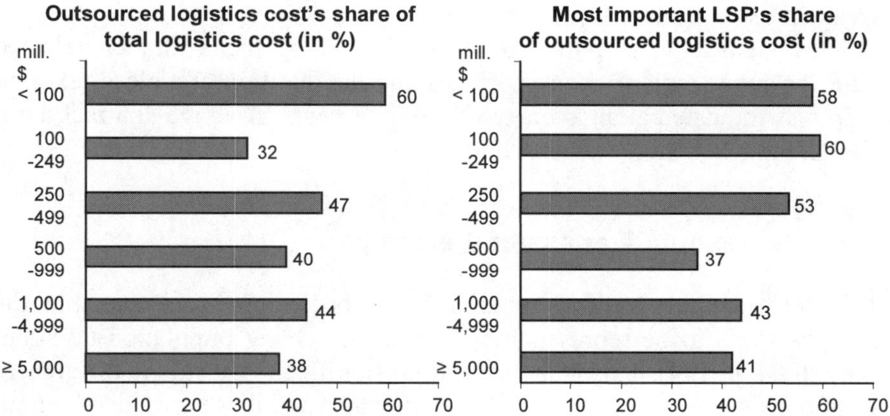

Fig. 5-3. Importance of logistics outsourcing (USA)

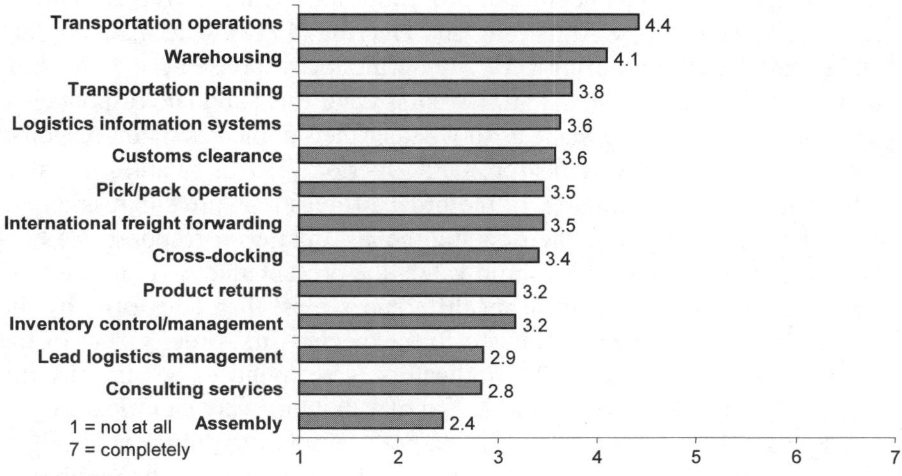

Fig. 5-4. Logistics services provided by the selected LSP (USA)

As in Germany, logistics outsourcing plays an important role in US firms, with an average of 42% spent on LSPs. Of this volume, an average of 45% is accounted for by the selected focal LSP, which is therefore responsible for a very large part of firms' logistics activities. The share of logistics cost spent on external service providers, as well as the focal LSP's share thereof are shown in Figure 5-3.

Considering outsourced logistics activities, firms in the US readily outsource transportation related services as well warehousing, while advanced activities are only outsourced by a small number of companies. Figure 5-4 graphically represents the extent to which logistics services are provided by the most important LSP.

Survey participants in the USA have worked together with their selected LSP for an average of about 5.5 years and the duration of a close relationship was indicated at an average of nearly 4 years. Average contract duration in the US sample was stated at 3.5 years.

5.2.3 Representativeness of the sample

In order to be able to draw generalizeable conclusions from a sample, the sample needs to be representative of the underlying population (Wallenburg 2004, p. 134). This is the case, if no significant systematic differences exist between answers provided by participants and those provided by non-participants of the study. In case distinctions between these two groups are detected, a non-response bias is diagnosed and the sample cannot be called representative. The most common procedure for testing for representativeness was developed by Armstrong and Overton (1977), who maintain that those participants answering very late during the response period are similar to those not responding at all. By comparing early and late respondents, a prediction can thus be made as to whether the sample has a non-response bias or not. Following Armstrong and Overton (1977), samples are split into four equally sized parts, in the order of questionnaire submission. T-tests are used to compare the first and the last quarter of responses. As literature does not reveal a threshold value, the present study will declare the sample biased if more significant differences exist than suggested by the significance level of the t-test, i.e. the expected first-order error. In the German sample, 14 out of 212 indicators were found to be significantly different at the 10%-level, while 21 out of 180 items were significantly different at the 10%-level in the USA. In both samples, the number of differences found at the 10%-level does not significantly exceed the number of differences expected to be found as first-order errors and the samples therefore do not appear to have a non-response bias.

Obtaining information from a single informant from an organization can lead to a systematic measurement error, called informant bias (see Ernst 2003). In order to rule out informant bias for the two samples, informant competency, as proposed by Kumar, Stern, and Anderson (1993, p. 1645), was assessed by examining respondents' characteristics. In the German (US) sample, 82.2% (96.8%) of respondents hold management positions in logistics, have been in their position for an average of 5.6 (5.6) years, and have worked for the same company for 10.8 (9.1) years. Hence, both samples' respondents are highly qualified for completing the questionnaire, as they are logistics managers and do dispose of sufficient organizational and functional experience. The samples can therefore not be assumed to be informant biased.

5.3 Introduction to the data analysis methodology

This section will introduce SEM as the methodology with which the collected data will be analyzed. SEM is a multivariate technique that infers dependencies between latent variables, or constructs, on the basis of empirically measured variances and covariances of observed variables, or indicators (Blalock 1963, p. 53). While SEM allows the simultaneous estimation of measurement and structural models, the two-step approach proposed by Anderson and Gerbing (1988) is the established procedure for applying SEM. In a first step, confirmatory factor analysis is employed to estimate the measurement models, while simultaneous estimation of measurement and structural models is conducted in a second step (Anderson and Gerbing 1988, p. 411). Section 5.3.1 will therefore provide an introduction to the basic concept of measurement models, while section 5.3.2 will outline the basic principles of structural models.

5.3.1 Measurement models

Bagozzi and Fornell (1982, p. 24) define a theoretical construct as "an abstract entity which represents the 'true', non-observable state or nature of a phenomenon." In order to measure such a latent construct, an understanding of the essential aspects and facets of the construct is necessary. On the basis of the conceptualization of the construct, a set of indicators to reflect the latent variable is defined and data is collected in a survey. This set of items is called a measurement model, or scale.

In general, two types of constructs can be distinguished: In first-order constructs, the whole set of indicators belonging to one measurement model can be aggregated into one construct, while indicators in multi-order constructs load onto more than one construct (Anderson, Gerbing, and

Hunter 1987, p. 435). Since the present study only uses first-order constructs, an explanation of multi-order constructs is not necessary.

For measuring a construct, one or more indicators can be operationalized, depending on the textual complexity of the construct. While single-item constructs are only capable of capturing the most simple issues, multi-item measures are generally preferred, as they are able to conceive more complex constructs (Churchill 1979, p. 66). In addition, two types of indicators can be distinguished. Formative indicators "are observed variables that are assumed to cause a latent variable" (Diamantopoulos and Winklhofer 2001, p. 269), while reflective indicators operate reversely by being effected by the latent variable. In this way, reflective indicators result from the existence of the latent variable, as shown in Figure 5-5, in which rectangles stand for (observed) indicators and ovals stand for latent variables (constructs). The study at hand exclusively uses reflective indicators, because these are applicable for measuring the proposed issues.

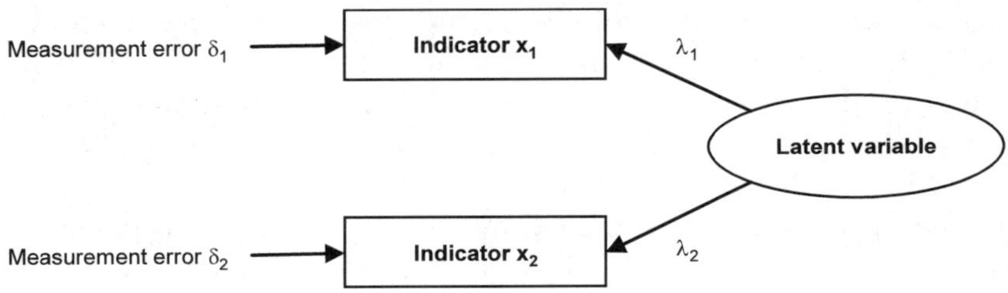

Fig. 5-5. Measurement model with reflective indicators

Confirmatory factor analysis (CFA) is the procedure conducted for construct measurement. Without going into mathematical detail, which can be found e.g. in Jöreskog and Sörbom (1982), CFA estimates model parameters by expressing the construct with its indicators as a linear system of equations and resembling the covariance matrix resulting from the model with the empirical covariance matrix as closely as possible. As a result of using reflective indicators, measurement is not free of error (Homburg and Giering 1996, p. 6) and a measurement error δ is estimated for each indicator, in addition to path coefficients λ between indicators and latent variables.

In order to be able to solve a CFA model, it has to be identified (Chou and Bentler 1995, p. 39) and for using congeneric models, as in the present study, a minimum of three indicators per construct is needed (Wallenburg 2004, p. 137). Following established procedure in SEM, all coefficients between observed variables and their measurement errors, as well as one coefficient between observed variables and latent variable per measurement model are set to one.

5.3.2 Structural models

A complete SEM model consists of at least two measurement models, which are put into relation in a structural model. While the measurement models define relations between observed variables and latent variables, the structural model defines the relations between latent variables (Bagozzi 1998, p. 67), i.e. independent constructs and dependent constructs, where the latter are being influenced by the former and where the strength of the relation is expressed by γ. While the algebraic signs '+' or '-' denote the direction of the relationship, its value indicates the absolute strength of the relationship. A sensible interpretation of γ, however, is only possible, if it is standardized.

Corresponding to the procedure described above for CFA, the SEM model is expressed as a linear equation system. Again, all parameters, now including measurement models, are estimated simultaneously, by converging the empirical covariance matrix with the covariance matrix expressed by the model. As the prediction of each of the dependent factors from the independent factors cannot be presumed to be without error, a residual error ζ is added to each dependent factor. A generic SEM model is displayed in Figure 5-6.

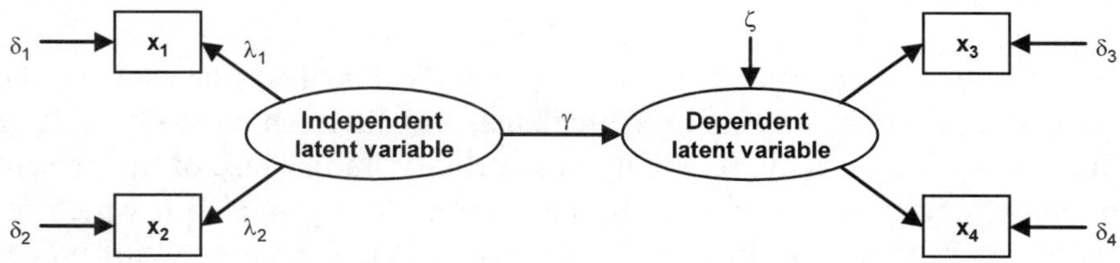

Fig. 5-6. Generic structural model

The quality of parameter estimation is the higher, the lower the difference between the two covariance matrices is. For minimizing this difference, a variety of estimation methods, including maximum likelihood (ML), generalized least squares, weighted least squares, or unweighted least squares can be used, which differ mainly in their assumptions about probability distribution, scale properties, and requirements regarding sample size (Golob 2003). The most common of these methods is ML, which is also used for the present study. Byrne (2001, p. 70) points out that the following conditions have to be met:

- A general minimum sample size cannot be established, as researchers generally propose different thresholds, e.g. a ratio of sample size to the number of free parameters of 5:1 (Bentler and Chou 1987, p. 91), but overall, researchers agree that sample size has to be fairly large.

- Multivariate normal distribution of data.
- Continuous scale of observed variables.

However, an advantage of ML estimation is its robustness against violations of these conditions. Bentler and Chou (1987, p. 89) state that "ML estimators are almost always acceptable even when data are nonnormally distributed" and Golob (2003, p. 9) comments on ML estimation's robustness when dealing with ordinal scales, such as the Likert scales used in the present research.

In any case, reliable estimations are only obtained when the SEM model is identified, i.e. when there are at least as many equations in the model as there are parameters to be estimated (Bagozzi and Baumgartner 1994, p. 390). However, a "just-identified model is not scientifically interesting because it has no degrees of freedom and therefore can never be rejected" (Byrne 2001, p. 35). Without degrees of freedom, several fit criteria cannot be calculated, which will be described in detail in the following section.

5.4 Model assessment

The quality of conclusions drawn from statistical analyses depends on the goodness of the underlying measurements, and attention has to be paid to the issues of objectivity, reliability, and validity. In the case of the present research, objectivity is ensured by the written survey mode, in which respondents cannot be influenced by the researcher. A wide array of procedures, however, has been developed for the assessment of reliability and validity. After briefly defining these two aspects, procedures for their assessment will be introduced.

5.4.1 Reliability

According to John and Reve (1982, p. 520), "reliability means low measurement error and indicates the extent to which measurements are repeatable and stable". While some indication of reliability of measurements in the present study can thus be obtained by comparing results to those obtained by Wallenburg (2004), where a high degree of similarity can be attested, internal consistency is of high importance. For this reason, Peter and Churchill Jr. (1986, p. 4) concretize the notion of reliability and define it as "the degree to which measures are free from random error and thus reliability coefficients estimate the amount of systematic variance in a measure". Hence, a respondent's score (X_0) consists of three components: the 'real' score (X_T), systematic error (E_S), and random error (E_R), as represented by the following equation:

$$X_0 = X_T + E_S + E_R \tag{1}$$

While systematic error does not diminish reliability, random error does, and following Homburg and Giering (1996, p. 6), reliability can be assumed when the individual indicators in a measurement model are highly correlated, such that the measurement model is homogeneous (Peter 1979, p. 8).

5.4.2 Validity

While reliability does not assess whether a measurement instrument does in fact measure what it is intended to, a measurement instrument is valid, "when the differences in observed scores reflect the true differences on the characteristic one is attempting to measure and nothing else, that is $X_0 = X_T$" (Churchill 1979, p. 65), such that the measurement is error-free. Accordingly, reliability is a necessary condition for validity, as it requires $E_R = 0$, while validity requires $E_S = E_R = 0$ (Peter and Churchill Jr. 1986, pp. 4-5).

While Garver and Mentzer (1999, p. 34) state a variety of facets of the notion of validity, only the following four aspects are relevant for the purposes of the research at hand (Wallenburg 2004, p. 142): content validity, convergent validity, discriminant validity, and nomological validity.

"Content validity exists when the scope of the construct is adequately reflected by the items as a group." (Dunn, Seaker, and Waller 1994, p. 157) Pertaining to making a judgment on the textual fit of items to the underlying construct, the assessment of content validity necessarily is of a qualitative nature (Parasuraman, Zeithaml, and Berry 1988, p. 28). Therefore, content validity can be assured by a careful and precise development of the measurement models and must be considered, when making model respecifications, which should never be grounded on statistical reasoning alone, as Anderson and Gerbing (1988, p. 416) stress.

Bagozzi and Phillips (1982, p. 468) define convergent validity as "the degree to which two or more attempts to measure the same concept through maximally dissimilar methods are in agreement." Convergent validity thus exists when indicators belonging to one construct have a high correlation with each other (Peter 1981, p. 136), which is examined when conducting an exploratory factor analysis (Homburg and Giering 1996, p. 8).

Discriminant validity refers to the degree to which measurements of different concepts are distinct (Bagozzi 1980, p. 130, Bagozzi 1981, p. 325). "The notion is that if two or more concepts are unique, then valid measures of each should not correlate too highly." (Bagozzi, Yi, and Phillips 1991, p. 425)

Finally, nomological validity refers to the relationships between constructs and is present if the hypothesized relationships between constructs can be validated on the basis of the data and is therefore assessed through an "explicit investigation of constructs and measures in terms of formal hypotheses derived from theory." (Peter and Churchill Jr. 1986, p. 5)

5.4.3 First generation criteria

First generation criteria are used to assess the validity and reliability of measurement instruments only. The three most important of these criteria, exploratory factor analysis, Cronbach's alpha, and adjusted item-total correlation, labeled traditional by Gerbing and Anderson (1988, p. 186), are briefly outlined below. While the former evaluates validity, the latter two assess reliability.

Exploratory factor analysis (EFA), like CFA, is based on the idea "that some underlying factors, which are smaller in number than the number of observed variables, are responsible for the covariation among the observed variables." (Dunn, Seaker, and Waller 1994, p. 161) Therefore, it is the aim of EFA to represent the entirety of indicators through as few factors as possible (see Mulaik 1987 for an examination of the history of EFA). In contrast to CFA, EFA is unrestricted, i.e. the structure of indicators is not fixed a priori through hypotheses. On the basis of correlations among indicators, factor loadings are calculated that express the association of items with corresponding factors. Factors are extracted through principal axis factoring and have to be rotated for the purpose of interpretation. While several rotation procedures are available (see Gorsuch 1997 and Browne 2001 for an introduction to the concept of rotation and a description of different procedures), the oblique technique of oblimin is used here. The criterion for determining the number of factors to be extracted was developed by Kaiser (1974), who proposed to only consider factors demonstrating high explanatory power through an Eigen-value above one.

On the basis of EFA, convergent and discriminant validity can be assessed preliminarily. Although general agreement on a minimum factor loading has not been reached, all indicators have to load unambiguously on only one factor, with merely minor associations to other factors. As suggested by Homburg and Giering (1998, pp. 128-129), the present study assumes convergent and discriminant validity at factor loadings exceeding 0.4, with an explained variance of over 50%.

Garver and Mentzer (1999, p. 43) state that "coefficient alpha is the most commonly used index of scale reliability." Introduced by Cronbach (1951), and therefore often referred to as Cronbach's alpha, coefficient alpha is used to appraise internal consistency reliability. In algebraic representation, coefficient alpha can be stated as (Cronbach 1951, p. 304):

$$\alpha = \frac{I}{I-1}\left(\frac{\sum_{i=1}^{I}\sigma_i^2}{\sigma_t^2}\right), \tag{2}$$

with I standing for the number of indicators in a measurement instrument, σ_i^2 standing for the variance of the ith indicator, and σ_t^2 standing for the variance of the measurement instrument. "Thus, a low coefficient alpha indicates the sample of items performs poorly in capturing a construct" (Churchill 1979, p. 68). While again a minimum coefficient alpha has not been established, sufficient internal consistency reliability can be assumed at an alpha of over 0.9 for established scales, while 0.5 or 0.6 can be sufficient for early research stages (Nunnally 1967, p. 226). Attention, however, has to be paid to the fact that coefficient alpha increases with the number of indicators included in the measurement (Garver and Mentzer 1999, p. 43), wherefore higher alphas have to be required for factors incorporating a large number of indicators, while lower alphas are tolerable for 'small' factors. For this thesis, in which most latent variables are composed of no more than four indicators, a coefficient alpha of 0.8 shall be sufficient. Since "Cronbach's Alpha can be looked upon as an average correlation of every combination of one question to the other questions in the group" (Mentzer and Flint 1997, p. 210), there have to be at least three indicators in a measurement instrument to be able to calculate coefficient alpha.

The last first generation criterion presented here is adjusted item-total correlation[5], which "is the statistical correlation between a given item and the scale to which it belongs. If a scale contains an item with a low item-to-scale correlation, then the scale may be deemed to be unreliable in that the items are not internally consistent." (Dunn, Seaker, and Waller 1994, p. 160) In this respect, adjusted item-total correlation is related to coefficient alpha, which can be improved by eliminating items with low item-total correlation, a procedure suggested by Churchill (1979, p. 68). A minimum value for adjusted item-total correlation does not have to be provided, as item-total correlations will only be employed in model assessment when coefficient alpha is insufficient.

5.4.4 Second generation criteria

First generation goodness of fit criteria, as pointed out in the previous section, take an exploratory approach to scale assessment. In contrast to this, confirmatory factor analysis (CFA), a special case of SEM, intends to assess a priori hypothesized factorial structures and a multitude of criteria for judging model adaptation has been developed on this basis. The most important ones, according to the categorization made by Homburg and Baumgartner (1995, p. 165), will be presented below (see Figure 5-7 for an overview).

5.4.4.1 Global adaptation measures

Global adaptation measures assess the adaptation of the model as a whole, i.e. how well the sample covariance matrix S and the model's covariance matrix $\Sigma(\theta)$ correspond (Chou and Bentler 1995, p. 44). Global adaptation measures can be subdivided into relative measures and measures with comparison level, depending on whether or not comparison levels exist.

Global adaptation measures with comparison level can be subdivided into stand-alone measures and incremental measures. While incremental measures assess adaptation in comparison to a baseline or null model,

[5] The terminology applied in research is fuzzy with regard to item-total correlation and several expressions (e.g. item-to-total correlation, item-to-scale correlation) are used to denote item-total correlation. In addition, both Wallenburg 2004, p. 146 and Willauer 2005, p. 163 maintain that two textual types of item-total correlation have to be distinguished: Simple item-total correlation, which is the correlation of an item with its scale including itself, and adjusted item-total correlation as an item's correlation with all items from a scale but itself. In the context of the present study, only adjusted item-total correlation is used.

stand-alone measures suffice without and can be further subdivided into descriptive measures and inference statistical measures.

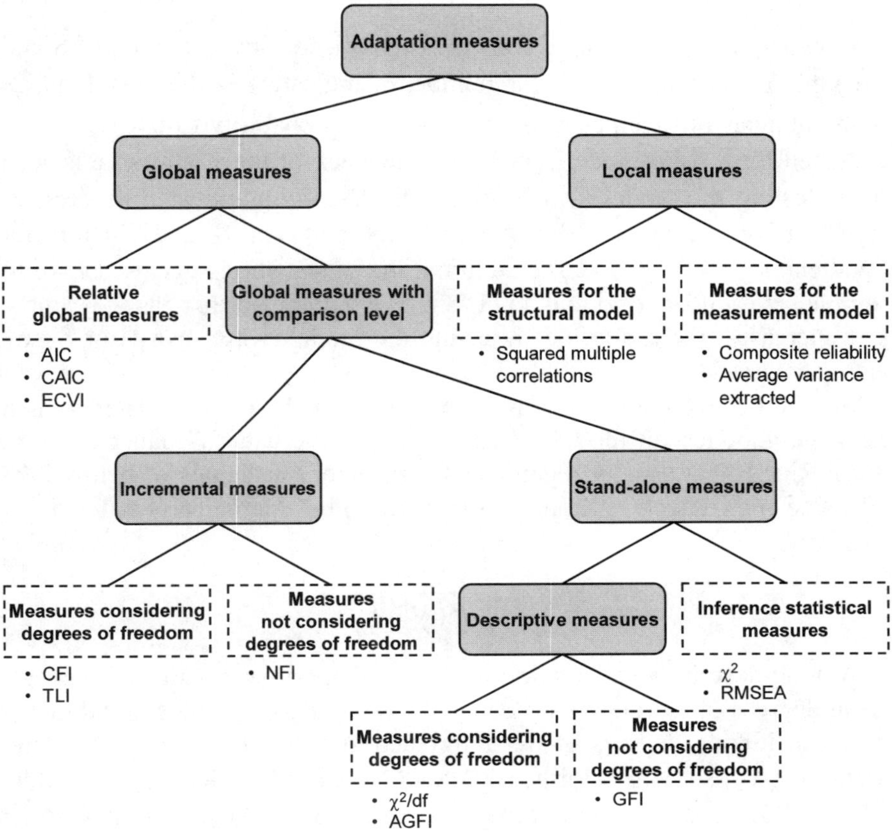

Fig. 5-7. Overview of adaptation measures[6]

Inference statistical measures are based on statistical testing procedures and the χ^2 statistic and the RMSEA (Root Mean Squared Error of Approximation) are used in this study.

The χ^2 test is based on the null hypotheses that S equals $\Sigma(\theta)$, where

$$\chi^2 = (N-1)\, F[S, \Sigma(\theta)], \tag{3}$$

and a χ^2 distribution with,

[6] Adapted from Homburg and Baumgartner 1995, p. 165.

$$df = \frac{1}{2} q(q+1) - t, \tag{4}$$

in which N is the sample size, F is the discrepancy function of S and $\Sigma(\theta)$ that is minimized, q is the number of indicators in the model, and t is the number of parameters in the model. A probability of 5% is usually expected for a valid model. The main drawback of the χ^2 measure is seen in its testing the model as a whole, such that in complicated models, χ^2 significantly increases, even though only parts of S and $\Sigma(\theta)$ differ (Wallenburg 2004, p. 148). In addition, the χ^2 statistic is very sensitive to sample size and Bagozzi and Yi (1988, p. 77) maintain that "as the sample size increases, the chances of rejecting the model (whether true or false), also increases."

In order to overcome the drawbacks of χ^2, RMSEA is calculated, which tests the goodness of model approximation to the data (Willauer 2005, p. 168). RMSEA is usually regarded as satisfactory at values of below 0.08 (Browne and Cudeck 1993, pp. 136ff) and can be algebraically stated as:

$$RMSEA = \sqrt{\frac{\chi^2 - df}{df(N-1)}}. \tag{5}$$

Among descriptive adaptation measures, there are measures not considering degrees of freedom, such as GFI (Goodness of Fit Index), and measures considering degrees of freedom, such as AGFI (Adjusted GFI), and adjusted χ^2 (also called normed χ^2). GFI, proposed by Jöreskog and Sörbom (1982), assesses the discrepancy between S and $\Sigma(\theta)$ in terms "of the relative amount of variances and covariances in S that are accounted for by the implied model $\Sigma[(\theta)]$" (Hu and Bentler 1995, p. 86) and is best when S equals $\Sigma(\theta)$, such that GFI takes a value of one. Formally stated,

$$GFI = 1 - \frac{tr\left[\hat{\Sigma}^{-1} S - I\right]^2}{tr\left[\hat{\Sigma}^{-1} S\right]^2}, \tag{6}$$

where $\hat{\Sigma}^{-1}$ is the inverse of the implied covariance matrix, tr is the sum of diagonal elements of the matrix (trace), and I is the identity matrix.

By introducing a penalty term for model complexity, GFI is adjusted, such that

$$AGFI = 1 - \left[\frac{t(t+1)}{2df}\right] \frac{tr\left[\hat{\Sigma}^{-1} S - I\right]^2}{tr\left[\hat{\Sigma}^{-1} S\right]^2} . \qquad (7)$$

Through this adjustment of GFI with a penalty term for the number of parameters (t), GFI's main deficit is addressed, i.e. the possibility of improving GFI by merely adding parameters. AGFI therefore prefers parsimonious models over complicated models and for both indices, a minimum value of 0.9 is usually suggested to indicate good model fit (Bagozzi and Yi 1988, p. 79). A similar adjustment can be made for χ^2 in order to overcome its preference for complicated models. Dividing χ^2 by the degrees of freedom yields the adjusted χ^2 (χ^2/df), according to which parsimonious models are preferred over complicated models. Good model adaptation is usually attested at an adjusted χ^2 of below 2.5 (Bollen 1989), although a variety of values between 2 and 3 are suggested by different researchers (e.g. Homburg and Baumgartner 1995).

Apart from stand-alone measures, incremental measures can be used to assess structural models. These measures, introduced to SEM by Bentler and Bonett (1980), use a baseline model as a comparison level, which assumes all indicators to be independent. Two widely used indices in this field are CFI (Comparative Fit Index) and TLI (Tucker-Lewis Index). The TLI "measures parsimony by assessing the degrees of freedom from the proposed model to the degrees of freedom from the null model." (Garver and Mentzer 1999, p. 41) In contrast to TLI, which is very sensitive to sample size, overrejecting models at small sample sizes, CFI is quite robust with regard to sample size. Both indices yield values in the zero to one range and a model is the better, the higher its CFI and TLI indices are. Garver and Mentzer (1999, p. 41) propose the use of both indices in logistics research and recommend values of at least 0.9 for good model fit. The formal representation given by Hu and Bentler (1995, pp. 83-85) of TLI and CFI is the following:

$$TLI = \frac{\frac{\chi_B^2}{df_B} - \frac{\chi_M^2}{df_M}}{\frac{\chi_B^2}{df_B} - 1}, \tag{8}$$

$$CFI = 1 - \frac{\max\left[\chi_M^2 - df_M, 0\right]}{\max\left[\chi_M^2 - df_M, \left(\chi_B^2 - df_B\right), 0\right]}, \tag{9}$$

where χ_B^2 is the χ^2 statistic for the baseline model, χ_M^2 is the χ^2 statistic for the hypothesized model, df_B is the number of degrees of freedom of the baseline model, and df_M is the number of degrees of freedom of the hypothesized model.

As the last variety of global adaptation measures, relative adaptation measures are used to compare different models. As a meaningful interpretation of measures is only possible if measures from different models can be juxtaposed, relative adaptation measures cannot be employed for the assessment of a single model. For the present study, two relative global adaptation measures are utilized, CAIC (Consistent Akaike's Information Criterion) and ECVI (Expected Cross-Validation Index). Both CAIC[7] and ECVI[8] relate the χ^2 statistic to the number of parameters and sample size and the indices can be formally stated as:

$$CAIC = \chi_M^2 + t\left(1 + \ln N\right), \tag{10}$$

$$ECVI = \left(\frac{\chi^2}{N}\right) + \left(\frac{2t}{N}\right). \tag{11}$$

When comparing models, the model demonstrating the lowest values for CAIC and ECVI is preferred over others.

[7] CAIC was developed by Bozdogan 1987 on the basis of Akaike's Information Criterion (AIC), see also Anderson, Burnham and White 1998 for a comparison of AIC and CAIC.

[8] ECVI was developed by Browne and Cudeck 1989, see also Reinecke 2003.

5.4.4.2 Local adaptation measures

While global adaptation measures properly address the overall fit of a model, they "do not explicitly provide information as to the nature of individual parameters and other aspects of the internal structure of a model." (Bagozzi and Yi 1988, p. 80) For this reason, a variety of local adaptation measures will be introduced for utilization in the present research. The first series (average variance extracted $AVE(\xi_j)$, composite reliability $\rho(\xi_j)$, individual item reliability $\rho(x_i)$, t-value of factor loadings, and Fornell/Larcker criterion) is used for the assessment of measurement models, while squared multiple correlations $R^2(\xi_j)$ are calculated for assessing the fit of structural models.

$AVE(\xi_j)$ refers to the average variance extracted from a measurement model ξ_j and measures, as stated by Fornell and Larcker (1981, p. 45), "the amount of variance that is captured by the construct in relation to the amount of variance due to measurement error." Bagozzi and Yi (1988, p. 82) recommend a minimum value of 0.5, as a lower value would signal that measurement accounts for a larger amount of variance than the factor itself. Formally stated,

$$AVE(\xi_j) = \frac{\sum_{i=1}^{q} \lambda_{ij}^2 \, \phi_{jj}}{\sum_{i=1}^{q} \lambda_{ij}^2 \, \phi_{jj} + \sum_{i=1}^{q} \theta_{ii}}, \tag{12}$$

where ξ_j is the latent variable j, λ_{ij} is the factor loading of item i belonging to the latent variable j, ϕ_{jj} is the variance of latent variable j, and θ_{ii} is the measurement error of item i.

Two important measures for examining the internal structure of a model are composite and individual item reliabilities (Bagozzi and Yi 1988, p. 80). Composite reliability $\rho(\xi_j)$ provides information about the internal consistency of a latent variable j. In contrast to that, individual item reliability $\rho(x_i)$ indicates to what extent the amount of variance of an indicator i is accounted for by the latent variable j. While both values are better, the larger the values are, Bagozzi and Yi (1988, p. 82) recommend a minimum value of 0.6 for composite reliability and Homburg and Baumgartner (1995, p. 170) indicate that a minimum of 0.4 is typically suggested for individual item reliability. The two measures can be formally stated as

$$\rho(\xi_j) = \frac{\left(\sum_{i=1}^{q} \lambda_{ij}\right)^2 \phi_{jj}}{\left(\sum_{i=1}^{q} \lambda_{ij}\right)^2 \phi_{jj} + \sum_{i=1}^{q} \theta_{ii}} \quad \text{and} \tag{13}$$

$$\rho(x_i) = \frac{\lambda_{ij}^2 \, \phi_{jj}}{\lambda_{ij}^2 \, \phi_{jj} + \theta_{ii}}. \tag{14}$$

Garver and Mentzer (1999, p. 43) demand that "each parameter estimate should be statistically significant", which can be assessed through t-values, i.e. "parameter estimates divided by their standard errors." (Bagozzi, Yi, and Phillips 1991, p. 431) Using a two-sided significance test, t-values of at least 1.96 indicate significant parameter estimates on the 5% level.

Discriminant validity can be examined using the Fornell/Larcker criterion. According to Fornell and Larcker (1981), a construct is discriminant, if the construct's average variance extracted is larger than the squared correlations with all other constructs.

Finally, squared multiple correlations $R^2(\xi_j)$ can be employed to assess the adaptation of structural models, as they indicate the proportion of the variance of latent variable j that is explained by the predictors of latent variable j (Byrne 2001, p. 163). A threshold value does not exist for $R^2(\xi_j)$, as the required explanatory power of a model depends on the research design, i.e. whether a model attempts partial explanation of latent variable j, or full explanation (Wallenburg 2004, p. 152). Algebraically, $R^2(\xi_j)$ can be stated as

$$R^2(\xi_j) = 1 - \frac{\psi_{jj}}{\phi_{jj}}, \tag{15}$$

where ψ_{jj} stands for the variance of the latent variable j's residual error.

5.4.5 Overview of goodness of fit criteria

In order to summarize the information provided in the preceding two sections, an overview of adaptation measures including recommended threshold values is given in Figure 5-8. It is critical to note, however, that not all adaptation measures have to be met in order to attest satisfactory model fit. Rather, evaluation and interpretation of results always has to take into consideration the theoretical fundament that a model is based upon. In this sense, several authors, including Fornell and Larcker (1981), Bagozzi and Yi (1988), and Homburg and Baumgartner (1995), point out that threshold values can only be seen as recommendations and that overall judgment of a model has to be based on the analysis of all available information, including both quantitative and qualitative reasoning.

	Adaptation measure	Threshold	
Fit of measurement models	Coefficient α	≥ 0.8	First generation criteria
	Explained factor variance	≥ 0.5	
	Factor loading	≥ 0.4	
	Adjusted item-total correlation	highest	
	t-value of factor loading	≥ 1.96	Second generation criteria
	Individual item reliability	≥ 0.4	
	Composite reliability	≥ 0.6	
	Average variance extracted	≥ 0.5	
Fit of measurement models and entire SEM model	CAIC	lowest	
	ECVI	lowest	
	Adjusted χ^2	≤ 2.5	
	TLI	≥ 0.9	
	GFI	≥ 0.9	
	AGFI	≥ 0.9	
	CFI	≥ 0.9	
	RMSEA	≤ 0.08	
Fit of structural models	Squared multiple correlations		

Fig. 5-8. Overview of fit criteria[9]

[9] Adapted from Wallenburg 2004, p. 153.

5.5 Model design and modification

5.5.1 Measurement models

A final measurement model has to be both reliable and valid in capturing the underlying construct. For this reason, the set of indicators gathered in data collection is subjected to a number of statistical tests to assure compliance with the fit criteria presented in the preceding sections. For the purposes of the present research, the procedure recommended by Homburg and Giering (1996) and refined by Wallenburg and Weber (2005, pp. 175-176) is applied, in which the set of indicators is examined first by using first generation criteria, and second by using second generation criteria.

Initially, coefficient alpha is to be calculated to assure the existence of internal consistency reliability. If coefficient alpha is below the recommended threshold of 0.8, the indicator displaying the lowest adjusted item-total correlation is to be eliminated. This procedure is proposed to be repeated until the threshold value is met. Next, an EFA is to be conducted with the aim of validating the single-factor structure of the measurement model. When only one factor is extracted, convergent validity can be attested. In addition, the factor is to explain at least 50% of the indicators' variance. In case more than one factor is extracted, or if explained factor variance is below 50%, those indicators are deleted that exhibit factor loadings of below 0.4.

The second step requires CFA and builds on the single-factor structure validated before. Here, global adaptation measures (CFI, TLI, RMSEA, GFI, AGFI, and adjusted χ^2)[10], as well as individual item and composite reliability are examined. Should several criteria not be met, indicators are to be eliminated on the basis of individual item reliabilities, such that indicators with reliabilities of below 0.4 are deleted. In the end, a reliable and convergent-valid measurement model is reached.

Wallenburg (2004, p. 155) maintains that the main advantage of this procedure is its high degree of rigor. However, he also points out that textual considerations are not included. In fact, Mulaik and James (1995, pp. 135-136) warn that "[t]his should be done cautiously and accompanied with explicit reasons" – merely modifying measurement models for statistical reasons does not endorse issues of content and nomological validity. Consequently, the present research will explicitly incorporate textual issues when considering model modification.

---

[10] Computation of these measures, however, requires at least one degree of freedom and can therefore only be calculated for measurement models including at least four indicators.

In addition, Wallenburg (2004, p. 155) points out that the sequential approach proposed by Homburg and Giering (1996) bears the danger of increasing convergent validity alone, when eliminating indicators for reasons of low adjusted item-total correlations, yet disobeying issues of reliability. As pointed out before, though, reliability is a necessary condition for validity. For this reason, both steps, i.e. application of first and second generation criteria, will be carried out simultaneously, such that reliability and validity issues are always considered jointly.

5.5.2 Structural models

SEM is geared at the examination of relationships between latent variables. Wallenburg (2004, p. 155), however, points out that this is only possible, if the SEM model meets certain fit criteria, i.e. the adaptation of the hypothesized model to the empirical data has to be satisfactory. Reasons for insufficient model adaptation can be attributed to bad quality of empirical data, inappropriate modeling of relationships, or overly complex model structure.

Fit criteria are necessarily the worse, the less identical the hypothesized and the empirical models are. Should a model not include relationships that are actually present in the empirical data, model fit can be improved by adding those relationships. An indication of this is provided in the form of modification indices, which state the change in the χ^2 statistic for each additional relationship (Byrne 2001, pp. 90-91). However, increasing model fit in this manner is not necessarily useful, as the analysis' confirmatory character is lost. In fact, Wallenburg (2004, p. 156) advises caution, as such modification may result in an over-specified model, which may not be generalizeable anymore.

Insufficient model fit may also result from an overly complex model. Since model complexity increases disproportionately to the number of constructs included, with each construct inducing multiple relationships to other constructs, model fit can be improved by eliminating constructs (Kaplan 1995, p. 100). This procedure, however, is only advisable, if the specific construct does not contribute explanatory value to the dependent variable(s) the model intends to explain. For this reason, constructs should only be eliminated, if the following two conditions are met:

(1) the decrease of squared multiple correlations of the downstream dependent variables is very small
(2) and construct elimination yields considerably improved model fit.

Should the proposed measures not result in satisfactory model fit, the model in its entirety has to be rejected and hypotheses cannot be evaluated. Overall, Bagozzi and Yi (1988, p. 93) concisely state that "one must rely upon a judicious interpretation of statistical and nonstatistical rules-of-thumb along with conceptual and philosophical criteria." Therefore, this research will discuss any required model modification in detail and elaborate both on statistical and textual considerations.

5.6 Comparative analysis

Comparative analyses are conducted using the multi-group analysis function of the AMOS software package to examine invariance between two samples. While two samples already exist for the analysis of national culture's influence on the customer loyalty model, with one data set each for Germany and the USA, the remaining moderating analyses require samples to be split before proceeding to the actual analyses. If moderators are latent variables, as opposed to single indicators, a single score for the moderator first has to be obtained. This is achieved by calculating an average score for each case, where rotated factor loadings are used to weight the individual scores. Then, both for single item and latent variable measurements, data sets are split. For that, two approaches are generally feasible: median splitting and a split into three parts, where the middle section is not considered in subsequent comparative analyses. The latter approach, however, is not suitable for application in this study, as subdivided data sets would not meet minimum sample-size requirements of SEM. Therefore, median splitting is conducted to separate the data set (Durvasula et al. 1993, Homburg, Giering, and Menon 2003). The resulting two data sets, one for high and another one for low values of the moderator, are used for multi-group analyses in AMOS.

The underlying idea of multi-group analysis is the examination of sample invariance (Steenkamp and Baumgartner 1998, pp. 80-81, Steenkamp and Baumgartner 1998, p. 406). For testing the moderating effects hypothesized in this study, invariance is only examined for the structural weights directed at customer loyalty. As no grounded reasoning for other possible sources of differences (variances, covariances, errors) exists, these are not considered in the following analyses and will always be estimated freely, i.e. separately for the two samples.

The procedure used here is based on Byrne (2001, pp.176-197), who suggests testing the fundamental hypothesis that both samples' covariance matrices are different, i.e.:

$$H_0: \ \Sigma(\theta_1) \neq \Sigma(\theta_2), \tag{16}$$

where $\Sigma(\theta_1)$ is the model covariance matrix in sub-sample 1 and $\Sigma(\theta_2)$ is the model covariance matrix of sub-sample 2. This hypothesis is reflected in the so called unconstrained model, in which all model parameters are freely estimated – no equality constraints between the two samples are imposed – and both samples are thus calculated separately. Then, a second estimation is performed, in which some equality constraints are introduced. Figure 5-9 depicts a generic comparative structural model to exemplify the analysis procedure. Since effects on customer loyalty are in the focus of the present study, constraints for the structural paths leading to customer loyalty are imposed, corresponding to the γ-paths in Figure 5-9, such that $\gamma_{1\_1} = \gamma_{1\_2}$ and $\gamma_{2\_1} = \gamma_{2\_2}$. If differences in the model actually exist with respect to these constrained paths, model adaptation for the constrained model is significantly worse than for the unconstrained model.

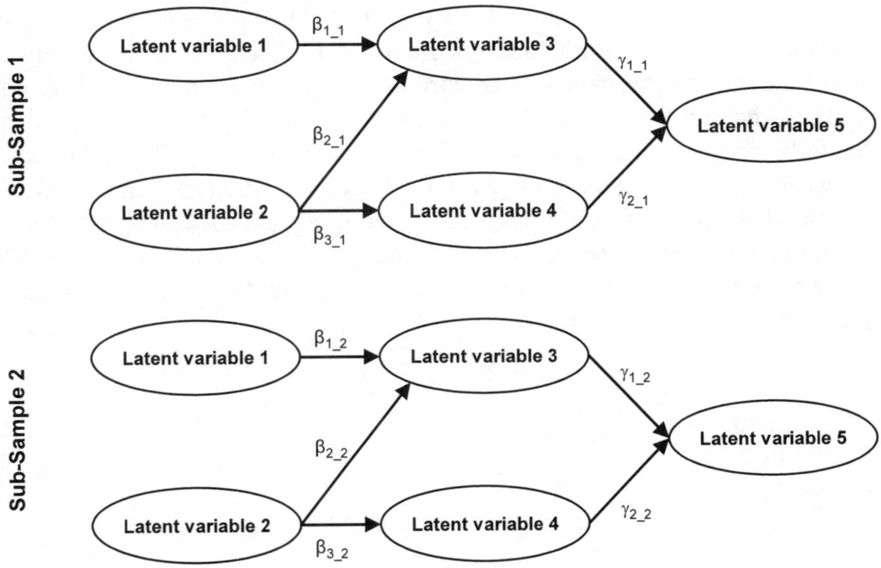

Fig. 5-9. Generic comparative structural model

In multi-group SEM, adaptation is measured through the χ^2 statistic (Granzin and Painter 2001, pp. 82-84) and the χ^2-difference ($\Delta\chi^2$) between the two models, as well as the difference in degrees of freedom (Δdf), are calculated, such that:

$$\Delta\chi^2 = \chi_C^2 - \chi_U^2, \quad \Delta df = df_C - df_U, \tag{17}$$

where χ_C^2 (χ_U^2) is the constrained (unconstrained) model's χ^2, df_C (df_U) is the constrained (unconstrained) model's degrees of freedom. The significance of $\Delta\chi^2$ with Δdf can be assessed employing the χ^2 distribution. In case the constrained model is significantly worse than the unconstrained model, differences with regard to the γ-paths exist between the two sub-samples – H_0 is supported – and further analyses have to be conducted to pinpoint the location(s) of the difference(s).

In order to find out which γ-path is different between the two sub-samples, they are individually examined in an iterative process (Thelen and Honeycutt Jr. 2004). For that, γ-paths are restricted to equality one-by-one, and the solution is compared to that of the previous less constrained model. If restricting a certain γ-path does not lead to a significant increase in χ^2, indicating that the path is not different between the sub-samples, the equality constraint is maintained for the following analyses. If χ^2 is significantly increased by imposing a constraint, the path is found to be different between the sub-samples and the respective γ-path is freely estimated in the following analyses. This procedure is conducted for all γ-paths contained in the model. Eventually, a model with all different γ-paths freely estimated and all invariant γ-paths restricted to equality is estimated. While this already reveals differences in the model, the individual parameter values for unequal γ-paths are also provided and allow the testing of directed hypotheses.

6 Operationalization and measurement

Chapter 4 provided conceptualizations of customer loyalty and its determinants and the methodology for measuring these concepts as constructs was described in chapter 5. The present chapter 6 deals with the operationalization of these constructs and their measurement in the empirical studies in Germany and the USA. As the present study is based on the empirical work of Wallenburg (2004), the following sections will introduce the measurement instruments as they are used by Wallenburg. If measurement instruments originate from other researchers, this will be indicated and reasons for modifications to Wallenburg's instruments, if necessary, will be given.

6.1 Customer loyalty dimensions

Wallenburg's examination of the empirical literature on customer loyalty reveals that established and comprehensively tested scales do not exist. Rather, most researchers have developed their own scales that are often tailored to a specific context (see Table 2-1), which can be attributed to the lack of a generally accepted customer loyalty conceptualization. In fact, most empirical studies capture customer loyalty as a one factorial construct, without differentiating between repurchases, additional purchases, and referrals. As this study is based on a three-dimensional conceptualization of customer loyalty, appropriate scales that do not require modifications, apart from Wallenburg's, were not identified.

6.1.1 Repurchases

The measurement of repurchases is guided by the operationalization of Giering (2000), which was developed for the industrial goods context and was modified by Wallenburg (2004, pp. 157-158) to account for the specific logistics outsourcing context (indicators 1-3). An additional indicator (Nr. 4) was introduced for this study to provide a more balanced representation of the concept. The indicators in their English version are shown in

Table 6-1. The full questionnaires in both German and English language are included in appendices A and B.

Table 6-1. Indicators for the measurement of "repurchase"

	Please indicate your level of agreement with the following statements regarding your past and future relationship with this LSP.
Indicator 1	We will continue using this LSP in the future.
Indicator 2	Right now, we intend to extend existing contracts with this LSP.
Indicator 3	If we knew then what we know now, we would again select this LSP.
Indicator 4	When the contract with this LSP ends, we will most likely negotiate a new contract with it, without starting a new bidding process.

Indicator 1 is a general measure capturing a customer's intention to continue its relationship with an LSP. Indicator 2 explicitly includes contract duration and indicator 4 was added in this study to provide more details on this by examining the mode in which contract extensions are typically conducted in the logistics field. In most cases, a new bidding process would be started at the end of a contract, in which a variety of LSPs may submit offers. If a customer is prepared to extend an existing contract without carrying out a new bidding process, this can be interpreted as a rather strong sign of loyalty. Indicator 3, finally, accounts for a customer's reflected considerations on the issue of repurchases. Assessment of the measurement model is shown in Table 6-2 for Germany and in Table 6-3 for the USA.

Table 6-2. Adaptation of "repurchase" (Germany, 4 indicators)

Information on the factor "repurchase" (Germany, 4 indicators)			
Coefficient alpha	0.81	AGFI	0.99
Explained factor variance	64.57%	CFI	1.00
χ^2/df	1.514	RMSEA	0.031
TLI	1.00	Composite reliability	0.82
GFI	1.00	Average variance extracted	0.53
Information on the indicators			
	Item-total correlation	*Individual item reliability*	*t-value of factor loadings*
Indicator 1	0.79	0.88	-
Indicator 2	0.76	0.78	31.83
Indicator 3	0.74	0.76	30.74
Indicator 4	0.39	0.16	9.71

Both in the USA and in Germany, indicator 4 does not exhibit satisfactory reliability. Since it only provides a specification of indicator 2, it can be eliminated from the measurement model without sacrificing content validity. The resulting three-indicator measurement corresponds to that of Wallenburg (2004), shows excellent model fit, and does not require further modification, as shown in Table 6-4 and Table 6-5.

Table 6-3. Adaptation of "repurchase" (USA, 4 indicators)

Information on the factor "repurchase" (USA, 4 indicators)			
Coefficient alpha	0.86	AGFI	0.99
Explained factor variance	64.38%	CFI	1.00
χ^2/df	0.281	RMSEA	0.000
TLI	1.01	Composite reliability	0.86
GFI	1.00	Average variance extracted	0.61
Information on the indicators			
	Item-total correlation	*Individual item reliability*	*t-value of factor loadings*
Indicator 1	0.84	0.91	-
Indicator 2	0.75	0.70	18.11
Indicator 3	0.72	0.63	16.62
Indicator 4	0.55	0.34	10.24

Table 6-4. Adaptation of "repurchase" (Germany, 3 indicators)

Information on the factor "repurchase" (Germany, 3 indicators)			
Coefficient alpha	0.92	AGFI	*
Explained factor variance	80.63%	CFI	*
χ^2/df	*	RMSEA	*
TLI	*	Composite reliability	0.92
GFI	*	Average variance extracted	0.80
* Value cannot be calculated, because factor with 3 indicators does not possess degrees of freedom.			
Information on the indicators			
	Item-total correlation	*Individual item reliability*	*t-value of factor loadings*
Indicator 1	0.88	0.89	-
Indicator 2	0.83	0.78	31.51
Indicator 3	0.83	0.75	30.58

Table 6-5. Adaptation of "repurchase" (USA, 3 indicators)

Information on the factor "repurchase" (USA, 3 indicators)			
Coefficient alpha	0.89	AGFI	*
Explained factor variance	74.51%	CFI	*
χ^2/df	*	RMSEA	*
TLI	*	Composite reliability	0.89
GFI	*	Average variance extracted	0.73
* Value cannot be calculated, because factor with 3 indicators does not possess degrees of freedom.			
Information on the indicators			
	Item-total correlation	*Individual item reliability*	*t-value of factor loadings*
Indicator 1	0.85	0.92	-
Indicator 2	0.77	0.69	17.05
Indicator 3	0.74	0.62	15.82

6.1.2 Additional purchases

Additional purchase intentions are even less researched than repurchase intentions and consequently, even fewer scales are available. Again, Wallenburg (2004, pp. 159-160) selected a scale developed by Giering (2000), which he modified to fit the logistics context. An additional indicator (Nr. 4) was introduced for the present study and the wording of the original three indicators was slightly modified to increase clarity.

Table 6-6. Indicators for the measurement of "additional purchase"

	Please indicate your level of agreement with the following statements regarding your past and future relationship with this LSP.
Indicator 1	In the future, the LSP will have a higher share of our logistics volume.
Indicator 2	In the future, we will use this LSP more than we do now.
Indicator 3	When we bid out other services than the ones we outsource today, we will consider this LSP preferentially.
Indicator 4	When we outsource additional services, we will first offer them to this LSP, before starting a bidding process.

Indicators 1 and 2 ask about the expected future extent of use of the currently most important LSP. Indicators 3 and 4 then refer to the way in which an LSP is selected for future projects. Table 6-7 and Table 6-8 show the fit of the measurement model in Germany and the USA.

Table 6-7. Adaptation of "additional purchase" (Germany, 4 indicators)

Information on the factor "additional purchase" (Germany, 4 indicators)			
Coefficient alpha	0.85	AGFI	0.73
Explained factor variance	60.64%	CFI	0.94
χ^2/df	31.679	RMSEA	0.237
TLI	0.83	Composite reliability	0.84
GFI	0.95	Average variance extracted	0.58
Information on the indicators			
	Item-total correlation	*Individual item reliability*	*t-value of factor loadings*
Indicator 1	0.65	0.62	-
Indicator 2	0.73	0.55	18.28
Indicator 3	0.59	0.38	14.70
Indicator 4	0.80	0.87	21.90

In both countries, fit of the measurement with 4 indicators is not satisfactory. In Germany, indicator 3, and in the USA indicator 4 are eliminated, because both indicators show unacceptable individual item reliabilities. Since indicators 3 and 4 display a considerable textual overlap, either one can be excluded without endangering content validity. The new measurement models consisting of 3 indicators each (see Table 6-9 and Table 6-10) show very good fit and do not require further modification.

Table 6-8. Adaptation of "additional purchase" (USA, 4 indicators)

Information on the factor "additional purchase" (USA, 4 indicators)			
Coefficient alpha	0.86	AGFI	0.53
Explained factor variance	64.32%	CFI	0.93
χ^2/df	28.750	RMSEA	0.334
TLI	0.78	Composite reliability	0.85
GFI	0.91	Average variance extracted	0.60
Information on the indicators			
	Item-total correlation	*Individual item reliability*	*t-value of factor loadings*
Indicator 1	0.77	0.90	-
Indicator 2	0.80	0.95	29.08
Indicator 3	0.75	0.48	14.06
Indicator 4	0.52	0.20	7.72

Table 6-9. Adaptation of "additional purchase" (Germany, 3 indicators)

Information on the factor "additional purchase" (Germany, 3 indicators)			
Coefficient alpha	0.85	AGFI	*
Explained factor variance	67.80%	CFI	*
χ^2/df	*	RMSEA	*
TLI	*	Composite reliability	0.86
GFI	*	Average variance extracted	0.67
* Value cannot be calculated, because factor with 3 indicators does not possess degrees of freedom.			
Information on the indicators			
	Item-total correlation	*Individual item reliability*	*t-value of factor loadings*
Indicator 1	0.72	0.63	-
Indicator 2	0.66	0.51	17.75
Indicator 4	0.80	0.89	19.93

Table 6-10. Adaptation of "additional purchase" (USA, 3 indicators)

Information on the factor "additional purchase" (USA, 3 indicators)			
Coefficient alpha	0.90	AGFI	*
Explained factor variance	77.53%	CFI	*
χ^2/df	*	RMSEA	*
TLI	*	Composite reliability	0.91
GFI	*	Average variance extracted	0.78
* Value cannot be calculated, because factor with 3 indicators does not possess degrees of freedom.			
Information on the indicators			
	Item-total correlation	*Individual item reliability*	*t-value of factor loadings*
Indicator 1	0.87	0.91	-
Indicator 2	0.88	0.95	27.38
Indicator 3	0.67	0.47	13.87

As a result of the textual overlap of indicators 3 and 4, differences in content between the German and US measurement models are very small and because the final measurement of customer loyalty in the USA differs considerably from Germany, as will be shown in section 6.1.4, a detailed discussion of possible differences is not necessary.

6.1.3 Referrals

Referrals can be made by customers at any time during the relationship, wherefore this construct can be measured by referring to actual referral behavior, instead of intentions. While the measurement model developed by Wallenburg (2004, p. 160) includes only one indicator (indicator 4), measurement was broadened in this thesis to provide a more accurate reflection of the concept.

Table 6-11. Indicators for the measurement of "referrals"

	Please indicate your level of agreement with the following statements regarding your past and future relationship with this LSP.
Indicator 1	Within our organization, we have recommended preferential consideration of this LSP for further projects.
Indicator 2	I often mention this LSP to my co-workers in a positive way.
Indicator 3	I often recommend this LSP to persons outside my company.
Indicator 4	We often recommend this LSP.

In addition to the original indicator 4, which captures general referral activities from the perspective of the respondent's organization, as expressed by the "we", indicator 1 specifies this organizational perspective by concretizing the subject of recommendations. In contrast, indicators 2 and 3 take a personal perspective, asking about individual recommendation behavior by the respondent. Here, referrals to other individuals within the respondent's organization (indicator 2) and to individuals outside the organization (indicator 3) are reflected.

Table 6-12. Adaptation of "referrals" (Germany, 4 indicators)

Information on the factor "referrals" (Germany, 4 indicators)			
Coefficient alpha	0.92	AGFI	0.76
Explained factor variance	75.09%	CFI	0.97
χ^2/df	28.570	RMSEA	0.225
TLI	0.92	Composite reliability	0.92
GFI	0.95	Average variance extracted	0.74
Information on the indicators			
	Item-total correlation	*Individual item reliability*	*t-value of factor loadings*
Indicator 1	0.68	0.44	-
Indicator 2	0.82	0.67	17.21
Indicator 3	0.88	0.95	19.78
Indicator 4	0.87	0.91	19.57

Table 6-13. Adaptation of "referrals" (USA, 4 indicators)

Information on the factor "referrals" (USA, 4 indicators)			
Coefficient alpha	0.94	AGFI	0.91
Explained factor variance	79.85%	CFI	0.99
χ^2/df	4.503	RMSEA	0.119
TLI	0.98	Composite reliability	0.94
GFI	0.98	Average variance extracted	0.80
Information on the indicators			
	Item-total correlation	*Individual item reliability*	*t-value of factor loadings*
Indicator 1	0.76	0.59	-
Indicator 2	0.85	0.74	15.13
Indicator 3	0.90	0.92	17.33
Indicator 4	0.91	0.93	17.52

Table 6-14. Adaptation of "referrals" (Germany, 3 indicators)

Information on the factor "referrals" (Germany, 3 indicators)			
Coefficient alpha	0.94	AGFI	*
Explained factor variance	84.10%	CFI	*
χ^2/df	*	RMSEA	*
TLI	*	Composite reliability	0.94
GFI	*	Average variance extracted	0.85
Information on the indicators			
	Item-total correlation	*Individual item reliability*	*t-value of factor loadings*
Indicator 2	0.79	0.65	-
Indicator 3	0.92	0.97	29.53
Indicator 4	0.90	0.90	28.65

Table 6-15. Adaptation of "referrals" (USA, 3 indicators)

Information on the factor "referrals" (USA, 3 indicators)			
Coefficient alpha	0.95	AGFI	*
Explained factor variance	86.20%	CFI	*
χ^2/df	*	RMSEA	*
TLI	*	Composite reliability	0.95
GFI	*	Average variance extracted	0.87
Information on the indicators			
	Item-total correlation	*Individual item reliability*	*t-value of factor loadings*
Indicator 2	0.84	0.73	-
Indicator 3	0.92	0.91	22.58
Indicator 4	0.92	0.94	23.01

Both in Germany (Table 6-12) and in the USA (Table 6-13), adjusted χ^2 and RMSEA do not meet the proposed standards. The reason for that, in both cases, is a high correlation between the error terms of indicators 1 and 2. As a result, indicator 1 was deleted for both countries, as it generally demonstrates inferior adaptation, as measured through item-total correlation and individual item reliability. From a textual point of view, both indicators are so interrelated that the elimination of either one still guarantees sufficient content validity. The resulting three-indicator measurements, with indicator 1 eliminated, shows excellent fit in both countries (Table 6-14 and Table 6-15).

6.1.4 Convergent and discriminant validity of customer loyalty

Although Wallenburg (2004, pp. 161-163) found the three dimensions of customer loyalty to be discriminant in his study in Germany, convergent and discriminant validity will again be assessed here. Especially with regard to the USA, where the three dimensional conceptualization of customer loyalty has not been validated before, confirmation of this approach has to be achieved. For this reason, two methods, as described in chapter 5.4, will be applied: EFA for testing for convergent validity, and the Fornell/Larcker criterion for assessing discriminant validity.

EFA on the German sample, as shown in Table 6-16, clearly indicates convergent validity for the three loyalty dimensions separately. In addition, Table 6-17 shows that the three dimensions of customer loyalty are discriminant, as each of the factors' average variance extracted is larger than the factors' correlations with each other. The Fornell/Larcker criterion is therefore satisfied.

Table 6-16. EFA on customer loyalty dimensions (Germany)

Indicator	Factor 1	Factor 2	Factor 3
Repurchase 1	0.939	0.442	0.570
Repurchase 2	0.886	0.409	0.478
Repurchase 3	0.870	0.402	0.573
Additional Purchase 1	0.363	0.797	0.410
Additional Purchase 2	0.426	0.722	0.525
Additional Purchase 4	0.395	0.942	0.460
Referrals 2	0.583	0.499	0.810
Referrals 3	0.555	0.512	0.985
Referrals 4	0.552	0.518	0.950

Table 6-17. Fornell/Larcker criterion for customer loyalty (Germany)

		Repurchase	Additional Purchase	Referrals
	AVE	0.92	0.67	0.85
Repurchase	0.92			
Additional Purchase	0.67	0.22		
Referrals	0.85	0.35	0.29	
AVE ≅ Average variance extracted				

In the USA, on the other hand, EFA does not support the three factorial structure of customer loyalty (Table 6-18), as all indicators are highly correlated with each other, i.e. they are convergent. Testing for discriminant validity in the USA, the Fornell/Larcker criterion (Table 6-19) supports existence of discriminance between the three constructs.

Table 6-18. EFA on customer loyalty dimensions (USA)[1]

Indicator	Factor 1	Factor 2	Factor 3
Repurchase 1	0.867		
Repurchase 2	0.805		
Repurchase 3	0.791		
Additional Purchase 1	0.779		
Additional Purchase 2	0.800		
Additional Purchase 3	0.797		
Referrals 2	0.848		
Referrals 3	0.848		
Referrals 4	0.892		

Table 6-19. Fornell/Larcker criterion for customer loyalty (USA)

		Repurchase	Additional Purchase	Referrals
	AVE	0.70	0.78	0.87
Repurchase	0.70			
Additional Purchase	0.78	0.58		
Referrals	0.87	0.69	0.45	
AVE ≅ Average variance extracted				

Since evidence on the dimensionality of customer loyalty is ambiguous in the USA, the hypothesis of a three dimensional construct cannot be supported, and a new measurement model has to be developed. For this purpose, the original 12 indicators of customer loyalty in the USA are taken as a starting point for the development of an integrated customer loyalty construct. All indicators together are included in Table 6-20.

Table 6-21 shows that the 12 indicator customer loyalty construct does not possess good fit, as most fit criteria are far from meeting the recommended threshold values. As a consequence, indicators 4 and 8 are eliminated because of their low individual item reliabilities.

[1] Only one factor is extracted, wherefore factor loadings for the other two factors cannot be reported.

Table 6-20. Indicators for the measurement of "loyalty" (USA only)

	Please indicate your level of agreement with the following statements regarding your past and future relationship with this LSP.
Indicator 1	We will continue using this LSP in the future.
Indicator 2	Right now, we intend to extend existing contracts with this LSP.
Indicator 3	If we knew then what we know now, we would again select this LSP.
Indicator 4	When the contract with this LSP ends, we will most likely negotiate a new contract with it, without starting a new bidding process.
Indicator 5	In the future, the LSP will have a higher share of our logistics volume.
Indicator 6	In the future, we will use this LSP more than we do now.
Indicator 7	When we bid out other services than the ones we outsource today, we will consider this LSP preferentially.
Indicator 8	When we outsource additional services, we will first offer them to this LSP, before starting a bidding process.
Indicator 9	Within our organization, we have recommended preferential consideration of this LSP for further projects.
Indicator 10	I often mention this LSP to my co-workers in a positive way.
Indicator 11	I often recommend this LSP to persons outside my company.
Indicator 12	We often recommend this LSP.

Table 6-21. Adaptation of "customer loyalty" (USA, 12 indicators)

Information on the factor "customer loyalty" (USA, 12 indicators)			
Coefficient alpha	0.95	AGFI	0.56
Explained factor variance	62.12%	CFI	0.79
χ^2/df	12.656	RMSEA	0.216
TLI	0.74	Composite reliability	0.95
GFI	0.70	Average variance extracted	0.60
Information on the indicators			
	Item-total correlation	*Individual item reliability*	*t-value of factor loadings*
Indicator 1	0.84	0.73	-
Indicator 2	0.79	0.63	15.78
Indicator 3	0.77	0.64	15.88
Indicator 4	0.57	0.32	9.69
Indicator 5	0.74	0.55	14.06
Indicator 6	0.76	0.58	14.67
Indicator 7	0.80	0.64	15.87
Indicator 8	0.55	0.29	9.26
Indicator 9	0.82	0.69	17.01
Indicator 10	0.81	0.74	18.09
Indicator 11	0.82	0.78	18.91
Indicator 12	0.85	0.83	20.03

Still, the model with 10 indicators shown in Table 6-22 fails to meet the required criteria. While individual item reliabilities do not suggest further indicators for elimination, high correlations between the error terms of indicators 5 and 6, and indicators 11 and 12 can be observed, leading to the elimination of indicators 5 and 11.

Table 6-22. Adaptation of "customer loyalty" (USA, 10 indicators)

Information on the factor "customer loyalty" (USA, 10 indicators)			
Coefficient alpha	0.96	AGFI	0.51
Explained factor variance	68.27%	CFI	0.79
χ^2/df	17.212	RMSEA	0.255
TLI	0.73	Composite reliability	0.95
GFI	0.69	Average variance extracted	0.68
Information on the indicators			
	Item-total correlation	*Individual item reliability*	*t-value of factor loadings*
Indicator 1	0.85	0.72	-
Indicator 2	0.79	0.62	15.35
Indicator 3	0.77	0.64	15.67
Indicator 5	0.75	0.54	13.84
Indicator 6	0.77	0.57	14.41
Indicator 7	0.78	0.62	15.41
Indicator 9	0.81	0.68	16.56
Indicator 10	0.83	0.76	18.15
Indicator 11	0.83	0.79	18.97
Indicator 12	0.87	0.84	20.09

Table 6-23. Adaptation of "customer loyalty" (USA, 8 indicators)

Information on the factor "customer loyalty" (USA, 8 indicators)			
Coefficient alpha	0.95	AGFI	0.80
Explained factor variance	68.47%	CFI	0.94
χ^2/df	5.994	RMSEA	0.142
TLI	0.92	Composite reliability	0.95
GFI	0.89	Average variance extracted	0.68
Information on the indicators			
	Item-total correlation	*Individual item reliability*	*t-value of factor loadings*
Indicator 1	0.85	0.77	-
Indicator 2	0.80	0.68	17.22
Indicator 3	0.77	0.65	16.63
Indicator 6	0.73	0.57	14.75
Indicator 7	0.78	0.64	16.25
Indicator 9	0.81	0.69	17.55
Indicator 10	0.82	0.72	18.18
Indicator 12	0.84	0.77	19.41

Although the elimination of indicators 5 and 11 has significantly improved model fit, several measures still do not satisfy the threshold values. Once more, high correlations between error terms can be observed, leading to the exclusion of indicators 1, 10, and 12, such that indicators 2, 3, 6, 7, and 9 remain in the model. Hence, the three perspectives of loyalty (repurchases, additional purchases, and referrals) are still represented.

The resulting 5 indicator measurement exhibits excellent model fit and does not require further modification. The comprehensive customer loyalty construct includes two indicators from each repurchases and additional purchases, and one indicator from referrals. It is interesting to note, though, that the new construct includes only the indicator from the original referrals construct, which captures the recommendation of preferential treatment of the LSP within the organization, neglecting referrals to individuals outside the respondent's organization. Comparing this to the German operationalization of referrals, and keeping cultural theory in mind, especially with regard to the research conducted by Hofstede (2001), the neglect of referrals to individuals outside the respondent's organization can be substantiated. It was pointed out before that Germans have a strong cohesion with their in-groups, a phenomenon underdeveloped in the USA. For this reason, Germans can be expected to be more inclined to sharing positive or negative experiences made with an LSP with other members of their in-groups, both inside and outside their organizations. In the USA, on the contrary, where different aspects of life are strictly separated, e.g. private and professional life spheres, specific knowledge and experiences made in one life sphere can be expected not to diffuse to other life spheres. Therefore, the referrals aspects not included in the integrated customer loyalty construct can be assessed not be important in the USA and the construct therefore appears sufficiently broad to capture all major facets of customer loyalty in the US context – content as well as nomological validity can be assumed. While the three dimensions of customer loyalty will continue to be used for all analyses encompassing German data, analyses on US data will employ the new integrated construct of customer loyalty.

Table 6-24. Adaptation of "customer loyalty" (USA, 5 indicators)

Information on the factor "customer loyalty" (USA, 5 indicators)			
Coefficient alpha	0.90	AGFI	0.96
Explained factor variance	65.02%	CFI	1.00
χ^2/df	1.510	RMSEA	0.045
TLI	0.99	Composite reliability	0.90
GFI	0.99	Average variance extracted	0.65
Information on the indicators			
	Item-total correlation	*Individual item reliability*	*t-value of factor loadings*
Indicator 2	0.80	0.72	-
Indicator 3	0.71	0.57	13.70
Indicator 6	0.72	0.58	13.98
Indicator 7	0.77	0.68	15.64
Indicator 9	0.78	0.69	15.83

6.2 Customer loyalty determinants

6.2.1 Satisfaction

Chapter 4.2.1 introduced the disconfirmation paradigm, which is the conceptual backbone of the assessment of satisfaction. For operationalizing satisfaction, it is important to determine, whether measurements should refer to the individual components of satisfaction according to the disconfirmation paradigm, or to an overall perception of satisfaction (Dabholkar 1995, p. 32). The former approach, i.e. measuring expectations and outcomes separately and assessing satisfaction by comparing these two components, is conceptually appealing. However, it bears some imprecision, because the role of different comparison levels for the evaluation of the actual outcome is not yet comprehensively researched (Wallenburg 2004, p. 163). For this reason, measurement instruments for satisfaction in this research, as in the preceding research by Wallenburg (2004), refer to an overall evaluation of the services provided by an LSP, asking respondents for their general perceptions on price satisfaction, service quality, and relational satisfaction. This is in line with the results achieved by Dabholkar, Shepherd, and Thorpe (2000), who compared the two approaches and found perception measures to be "clearly superior to computed disconfirmation on all criteria." (p. 167)

6.2.1.1 Price satisfaction

Price satisfaction is assessed as a result of evaluating all experiences made with an LSP regarding its prices. Individual transactions are not taken into consideration, because logistics outsourcing relationships typically have a long-term character.

Table 6-25. Indicators for the measurement of "price satisfaction"

	Please indicate your level of agreement with the following statements regarding your LSP's pricing.
Indicator 1	Compared to other LSPs, our LSP's prices are very good.
Indicator 2	The LSP offers a very good price-performance ratio.
Indicator 3	Compared to carrying out those tasks ourselves, our LSP's prices are very good.

Since he was not able to identify an appropriate scale in existing research, Wallenburg (2004, pp. 164-165) developed his own measurement instrument, which comprised two indicators[2]. These indicators differ in

[2] Simultaneous to Wallenburg's study, Siems 2003, pp. 106-108 developed a similar scale.

their comparison standard – while one indicator assesses price satisfaction in relation to prices offered by other LSPs, the second indicator compares the LSP's prices to the costs of carrying out the tasks in-house. In order to broaden the measurement, and to provide a measurement that does not only include price separately, but also in relation to the received services, a third indicator was introduced for the present research, assessing the respondent's perception of the achieved price-performance ratio. These three indicators are displayed in Table 6-25.

While the measurement model may be criticized for its reference to competing LSPs' prices, which may not be accurately known, this is of little importance with respect to the perception of price satisfaction. Even if respondents do not possess detailed knowledge of prices offered by other LSPs, whether or not they are satisfied with their current LSP's prices depends on a (subjective) estimate or evaluation of other LSPs' prices.

Table 6-26. Adaptation of "price satisfaction" (Germany, 3 indicators)

Information on the factor "price satisfaction" (Germany, 3 indicators)			
Coefficient alpha	0.74	AGFI	*
Explained factor variance	50.21%	CFI	*
χ^2/df	*	RMSEA	*
TLI	*	Composite reliability	0.74
GFI	*	Average variance extracted	0.49
* Value cannot be calculated, because factor with 3 indicators does not possess degrees of freedom.			
Information on the indicators			
	Item-total correlation	*Individual item reliability*	*t-value of factor loadings*
Indicator 1	0.59	0.56	-
Indicator 2	0.62	0.64	11.11
Indicator 3	0.48	0.30	10.64

Table 6-27. Adaptation of "price satisfaction" (USA, 3 indicators)

Information on the factor "price satisfaction" (USA, 3 indicators)			
Coefficient alpha	0.89	AGFI	*
Explained factor variance	73.23%	CFI	*
χ^2/df	*	RMSEA	*
TLI	*	Composite reliability	0.89
GFI	*	Average variance extracted	0.73
* Value cannot be calculated, because factor with 3 indicators does not possess degrees of freedom.			
Information on the indicators			
	Item-total correlation	*Individual item reliability*	*t-value of factor loadings*
Indicator 1	0.75	0.66	-
Indicator 2	0.84	0.89	15.82
Indicator 3	0.75	0.65	14.46

While the measurement model shows excellent fit for the USA, adaptation to the German data is not satisfactory. Both coefficient alpha and average variance extracted fall short of the recommended thresholds and explained factor variance just barely meets the requirements. Examining individual item reliabilities, it is obvious that indicator 3 is not closely associated with the factor, as expressed through a reliability value of 0.3, and is thus eliminated, to form the two-indicator measurement displayed in Table 6-28, which does not require further modification.

Table 6-28. Adaptation of "price satisfaction" (Germany, 2 indicators)

Information on the factor "price satisfaction" (Germany, 2 indicators)			
Coefficient alpha	**	AGFI	*
Explained factor variance	60.00%	CFI	*
χ^2/df	*	RMSEA	*
TLI	*	Composite reliability	**
GFI	*	Average variance extracted	**
* Value cannot be calculated, because factor with 3 indicators does not possess degrees of freedom.			
** CFA and coefficient alpha require at least three indicators.			
Information on the indicators			
	Item-total correlation	*Individual item reliability*	*t-value of factor loadings*
Indicator 1	0.60	**	**
Indicator 2	0.60	**	**

While a comparison of current LSP's prices with the costs of carrying out the respective logistics activities in-house forms part of a customer's price satisfaction in the USA, this aspect is apparently not considered in Germans' evaluations of prices. One possible explanation for this relates to Germans' fixation to the future, in contrast to Americans' focus on the present. Hence, decision making in Germany can be surmised to be directed at the future and consequently, decisions taken in the past that would be difficult or costly to reverse would only have a limited impact on supplier evaluations. Americans, on the contrary, do not attribute equal importance to the future, but concentrate on the present and on realizing immediate gains. Considering that Americans also do not value relationships as highly as Germans, the obtained results for price satisfaction can be explained. To Germans, in-sourcing a previously outsourced logistics activity may just not be a strategic option with relevance for the future, while it may be to Americans. In this sense, in-sourcing is not part of Germans' sets of alternatives, wherefore the costs of carrying out these tasks in-house do not affect assessments of an LSP's prices. Americans, on the other hand, can be expected to be more willing to impose change, which may lead to in-sourcing currently outsourced logistics activities. As in-sourcing may thus be an option to Americans, it is comprehensible that the costs of carrying out an outsourced logistics activity in-house form part of the evaluation of satisfaction with an LSP's prices in the USA.

6.2.1.2 Relational satisfaction

Relational satisfaction captures the customer's satisfaction with all aspects regarding the interaction and relationship with its most important LSP. As with price satisfaction, relational satisfaction is measured globally, grasping a respondent's evaluation of all experiences made in the relationship. Since Wallenburg (2004, pp. 165-166) was again not able to identify an appropriate scale in existing research, he developed a measurement instrument consisting of two indicators. One indicator refers to the interaction with the LSP specifically, while the other is more general in referring to the relationship in general. For the present research, a third indicator concerned with the way conflicts are handled was introduced, because dealing with differences is a very important aspect in relationships. Table 6-29 displays the measurement model. As can be seen in Table 6-30 and in Table 6-31, the measurement model has excellent fit in both the USA and Germany.

Table 6-29. Indicators for the measurement of "relational satisfaction"

	Please indicate your level of agreement with the following statements on how satisfied you are with the relationship between this logistics service provider and your company.
Indicator 1	We are very satisfied with the way we interact with our LSP.
Indicator 2	Differences when cooperating with this LSP are always settled smoothly.
Indicator 3	The relationship with this LSP is very good.

Table 6-30. Adaptation of "relational satisfaction" (Germany, 3 indicators)

Information on the factor "relational satisfaction" (Germany, 3 indicators)			
Coefficient alpha	0.91	AGFI	*
Explained factor variance	76.40%	CFI	*
χ^2/df	*	RMSEA	*
TLI	*	Composite reliability	0.91
GFI	*	Average variance extracted	0.76
* Value cannot be calculated, because factor with 3 indicators does not possess degrees of freedom.			

Information on the indicators			
	Item-total correlation	*Individual item reliability*	*t-value of factor loadings*
Indicator 1	0.80	0.74	-
Indicator 2	0.79	0.70	24.39
Indicator 3	0.84	0.85	26.86

6.2.1.3 Service quality

Service quality, conceptualized as the satisfaction with the services provided by an LSP, has been subject to empirical scrutiny by a multitude of researchers. Most commonly, service quality has been measured using the

SERVQUAL instrument, which was developed by Parasuraman, Zeithaml, and Berry (1985, 1988). With its five-dimensional structure, the SERVQUAL instrument measures tangibles, responsiveness, empathy, reliability, and assurance. For each of these dimensions, expected service quality and actual service quality are captured and the difference between expectations and actual outcomes marks confirmation or disconfirmation. On the one hand, as pointed out above, measurement of satisfaction by directly following the disconfirmation paradigm bears some difficulties. In addition, several researchers who tested the SERVQUAL instrument in different settings found that it is unapt in industrial service contexts (Bienstock, Mentzer, and Bird 1997, Mentzer, Flint, and Kent 1999) and Brown, Churchill Jr., and Peter (1993, p. 138) doubt that "a scale to measure service quality can be universally applicable across industries." Therefore, SERVQUAL is not employed in this research. Instead, the instrument developed by Dabholkar, Shepherd, and Thorpe (2000, p. 151) and modified by Wallenburg (2004, pp. 167-168) is used in this study.

Table 6-31. Adaptation of "relational satisfaction" (USA, 3 indicators)

Information on the factor "relational satisfaction" (USA, 3 indicators)			
Coefficient alpha	0.91	AGFI	*
Explained factor variance	77.53%	CFI	*
χ^2/df	*	RMSEA	*
TLI	*	Composite reliability	0.91
GFI	*	Average variance extracted	0.78
* Value cannot be calculated, because factor with 3 indicators does not possess degrees of freedom.			
Information on the indicators			
	Item-total correlation	*Individual item reliability*	*t-value of factor loadings*
Indicator 1	0.80	0.72	-
Indicator 2	0.80	0.73	16.92
Indicator 3	0.86	0.88	18.53

Table 6-32. Indicators for the measurement of "service quality"

	Please indicate your level of agreement with the following statements on your satisfaction with the LSP's service quality.
Indicator 1	Overall, this LSP offers excellent services.
Indicator 2	This LSP offers great performance.
Indicator 3	This LSP offers very high quality.

While indicators 1 and 3 were only slightly modified from the original Dabholkar, Shepherd, and Thorpe (2000) scale (referring to "LSP" instead of "company X"), indicator 2 refers to performance instead of quality, which assures a better fit to the logistics context. As provided in Table 6-33 and in Table 6-34, the measurement model has excellent fit in both Germany and the USA and does not require modification.

Table 6-33. Adaptation of "service quality" (Germany, 3 indicators)

Information on the factor "service quality" (Germany, 3 indicators)			
Coefficient alpha	0.95	AGFI	*
Explained factor variance	85.49%	CFI	*
χ^2/df	*	RMSEA	*
TLI	*	Composite reliability	0.95
GFI	*	Average variance extracted	0.86
* Value cannot be calculated, because factor with 3 indicators does not possess degrees of freedom.			
Information on the indicators			
	Item-total correlation	*Individual item reliability*	*t-value of factor loadings*
Indicator 1	0.88	0.84	-
Indicator 2	0.90	0.89	38.55
Indicator 3	0.88	0.83	35.15

Table 6-34. Adaptation of "service quality" (USA, 3 indicators)

Information on the factor "service quality" (USA, 3 indicators)			
Coefficient alpha	0.96	AGFI	*
Explained factor variance	88.24%	CFI	*
χ^2/df	*	RMSEA	*
TLI	*	Composite reliability	0.96
GFI	*	Average variance extracted	0.88
* Value cannot be calculated, because factor with 3 indicators does not possess degrees of freedom.			
Information on the indicators			
	Item-total correlation	*Individual item reliability*	*t-value of factor loadings*
Indicator 1	0.92	0.90	-
Indicator 2	0.91	0.88	29.44
Indicator 3	0.90	0.87	28.92

6.2.2 Proactive improvement

Proactive improvement is meant to measure the impetus with which an LSP continually improves its customer's logistics processes. While proactive improvement is an important factor in logistics outsourcing relationships, a scale was only developed by Engelbrecht (2004). Previous logistics research, as examined in the meta-analysis of scales used in logistics research by Keller et al. (2002), does not provide indication of a similar measurement instrument. For this reason, Wallenburg (2004, pp. 182-183) modified the scale used by Engelbrecht (2004) and proposed the indicators displayed in Table 6-35.

Indicator 5 is the most global of the five indicators, measuring the general innovativeness of the LSP. Indicator 1 is also rather general and captures customers' perceptions on LSPs' continuous improvement efforts. The remaining indicators are more specific. Indicator 2 refers to improvement efforts outside the LSP's direct responsibility, a very strong proof of

LSPs' motivation to increase their customers' performance. Indicators 3 and 4 finally include self-motivational aspects, i.e. they measure if an LSP strives to improve its customer's logistics processes, without direct motivation by the customer.

Table 6-35. Indicators for the measurement of "proactive improvement"

	Please indicate your level of agreement with the following statements on the continuous improvement efforts of this LSP.
Indicator 1	The LSP puts strong effort into continuously optimizing logistics processes.
Indicator 2	The LSP continuously makes suggestions for improvements of activities, even those outside its direct responsibility.
Indicator 3	When the situation changes, the LSP by itself modifies the logistics activities and processes, if this is useful and necessary.
Indicator 4	The LSP shows initiative by approaching us with suggestions for improvement.
Indicator 5	The LSP shows a high level of innovation.

Table 6-36. Adaptation of "proactive improvement" (Germany, 5 indicators)

Information on the factor "proactive improvement" (Germany, 5 indicators)			
Coefficient alpha	0.92	AGFI	0.97
Explained factor variance	69.92%	CFI	0.99
χ^2/df	3.178	RMSEA	0.063
TLI	0.99	Composite reliability	0.92
GFI	0.99	Average variance extracted	0.70
Information on the indicators			
	Item-total correlation	*Individual item reliability*	*t-value of factor loadings*
Indicator 1	0.75	0.61	-
Indicator 2	0.75	0.61	19.70
Indicator 3	0.80	0.70	21.48
Indicator 4	0.85	0.80	23.48
Indicator 5	0.83	0.77	22.80

Table 6-37. Adaptation of "proactive improvement" (USA, 5 indicators)

Information on the factor "proactive improvement" (USA, 5 indicators)			
Coefficient alpha	0.93	AGFI	0.92
Explained factor variance	74.00%	CFI	0.99
χ^2/df	3.688	RMSEA	0.104
TLI	0.98	Composite reliability	0.93
GFI	0.97	Average variance extracted	0.74
Information on the indicators			
	Item-total correlation	*Individual item reliability*	*t-value of factor loadings*
Indicator 1	0.83	0.72	-
Indicator 2	0.88	0.85	20.40
Indicator 3	0.64	0.43	11.74
Indicator 4	0.88	0.86	20.66
Indicator 5	0.86	0.82	19.71

As it can be seen in Table 6-36 for Germany and in Table 6-37 for the USA, the measurement models are nearly sufficient. Only adjusted χ^2 is slightly above the recommended threshold value in both countries, as well as RMSEA in the USA. While exceeding only one or two criteria does not necessarily induce modification of the model, in this case high correlations of the error terms of indicators 1 and 4 in Germany and the USA, and of indicators 1 and 3 in the USA, cause some disturbance in the models. Since indicator 1 is formulated as a general indicator, it can be eliminated without sacrificing content validity for the whole instrument, as the more specific indicators remain. Table 6-38 and Table 6-39 show that the elimination of indicator 1 improved model fit considerably.

Table 6-38. Adaptation of "proactive improvement" (Germany, 4 indicators)

Information on the factor "proactive improvement" (Germany, 4 indicators)			
Coefficient alpha	0.91	AGFI	0.99
Explained factor variance	71.95%	CFI	1.00
χ^2/df	0.704	RMSEA	0.000
TLI	1.00	Composite reliability	0.91
GFI	1.00	Average variance extracted	0.72
Information on the indicators			
	Item-total correlation	*Individual item reliability*	*t-value of factor loadings*
Indicator 2	0.74	0.60	-
Indicator 3	0.79	0.69	20.91
Indicator 4	0.85	0.83	23.34
Indicator 5	0.81	0.75	22.09

Table 6-39. Adaptation of "proactive improvement" (USA, 4 indicators)

Information on the factor "proactive improvement" (USA, 4 indicators)			
Coefficient alpha	0.91	AGFI	1.00
Explained factor variance	74.05%	CFI	1.00
χ^2/df	0.091	RMSEA	0.000
TLI	1.01	Composite reliability	0.92
GFI	1.00	Average variance extracted	0.74
Information on the indicators			
	Item-total correlation	*Individual item reliability*	*t-value of factor loadings*
Indicator 2	0.86	0.83	-
Indicator 3	0.63	0.42	12.08
Indicator 4	0.88	0.90	25.73
Indicator 5	0.85	0.82	23.07

6.2.3 Fairness

As pointed out in chapter 4.2.3, fairness measures the customer's perception of distributive justice or injustice in its relationship with an LSP, which results from an evaluation of the customer's input-output ratio. Consequently, two approaches for measuring fairness are feasible. On the one hand, both components of the evaluation, i.e. inputs and outputs, can be evaluated separately and be combined to assess fairness. On the other hand, an overall evaluation of perceived fairness can be taken. While the former approach is conceptually appealing, it has the disadvantage of inducing a considerable level of subjectivity to the judgment. As pointed out before with respect to the measurement of satisfaction, the role of comparison levels in such evaluations is not yet comprehensively researched, wherefore the second approach, i.e. obtaining an overall assessment on fairness within a relationship, is to be preferred.

While a multitude of logistics and marketing researchers incorporated social exchange theory in their empirical studies, equity theory is not commonly found. In fact, Wallenburg (2004) only identifies one scale for fairness that is appropriate in the customer loyalty and logistics context. This scale was developed by Moore and Cunningham III (1999) and modified and extended by Wallenburg (2004, pp. 183-185). While Wallenburg's instrument included 5 indicators, grasping perceptions on benefits sharing (indicator 1), self-interest of the LSP (indicator 2), fair treatment (indicator 3), balanced benefits for LSP and customer (indicator 4), and risk sharing (indicator 5), the last indicator is not used in the present research. It had to be excluded from Wallenburg's analyses, because its error term was highly correlated with the error term of indicator 4, indicating strong similarity of the two indicators. Examining indicator 5 from a content validity perspective, it does not add additional insight, although risk sharing is not explicitly included in the other indicators. However, indicators 1 and 4, which refer to the sharing of savings and benefits, are closely related to risk sharing and the remaining 4-indicator instrument (see Table 6-40) thus provides sufficient scope for the assessment of fairness.

Table 6-40. Indicators for the measurement of "fairness"

	Please indicate your level of agreement with the following statements on how satisfied you are with the relationship between this logistics service provider and your company.
Indicator 1	The LSP's own savings from process improvements are passed on to us to a fair extent.
Indicator 2	This LSP only looks out for itself.
Indicator 3	We feel fairly treated by this LSP.
Indicator 4	We and the LSP benefit from this outsourcing relationship to the same extent.

It is obvious in Table 6-41 that the measurement instrument does not possess adequate fit in Germany. Coefficient alpha, explained factor variance, as well as average variance extracted on a scale level, and the individual item reliabilities of indicators 1 and 2 do not meet the set standards. While both indicators 1 and 2 could be eliminated on the basis of their reliabilities, indicator 2 is excluded in a first step, because its error term additionally possesses a high correlation with indicator 4's error term.

In contrast, the measurement model in the USA, as depicted in Table 6-42, has satisfactory fit. Also, no significant correlations between error terms are observed, such that the model does not require modification.

Table 6-41. Adaptation of "fairness" (Germany, 4 indicators)

Information on the factor "fairness" (Germany, 4 indicators)			
Coefficient alpha	0.78	AGFI	0.96
Explained factor variance	48.69%	CFI	0.99
χ^2/df	2.143	RMSEA	0.078
TLI	0.97	Composite reliability	0.78
GFI	0.99	Average variance extracted	0.47
Information on the indicators			
	Item-total correlation	*Individual item reliability*	*t-value of factor loadings*
Indicator 1	0.53	0.36	-
Indicator 2	0.52	0.36	10.75
Indicator 3	0.67	0.67	12.62
Indicator 4	0.63	0.57	12.41

Table 6-42. Adaptation of "fairness" (USA, 4 indicators)

Information on the factor "fairness" (USA, 4 indicators)			
Coefficient alpha	0.82	AGFI	0.99
Explained factor variance	56.81%	CFI	1.00
χ^2/df	0.393	RMSEA	0.000
TLI	1.01	Composite reliability	0.82
GFI	1.00	Average variance extracted	0.54
Information on the indicators			
	Item-total correlation	*Individual item reliability*	*t-value of factor loadings*
Indicator 1	0.59	0.42	-
Indicator 2	0.59	0.41	8.58
Indicator 3	0.72	0.70	10.43
Indicator 4	0.59	0.74	10.52

Table 6-43 exhibits the modified instrument for fairness in Germany. Having eliminated indicator 2, the model is satisfactory, except for indicator 1's still low individual item reliability. Since everything else meets the standards, indicator 1 will nevertheless not be deleted in order not to endanger content validity.

The fact that indicator 2 in Germany reduces model fit, while it displays adequate fit in the USA can most likely be attributed to the question's wording. In the German questionnaire, the statement is phrased somewhat harder than in the US version. This results in a lower average agreement with the statement in Germany, which is 4.88 in comparison to 5.76 in the USA.

Table 6-43. Adaptation of "fairness" (Germany, 3 indicators)

Information on the factor "fairness" (Germany, 3 indicators)			
Coefficient alpha	0.75	AGFI	*
Explained factor variance	53.15%	CFI	*
χ^2/df	*	RMSEA	*
TLI	*	Composite reliability	0.76
GFI	*	Average variance extracted	0.51
* Value cannot be calculated, because factor with 3 indicators does not possess degrees of freedom.			
Information on the indicators			
	Item-total correlation	*Individual item reliability*	*t-value of factor loadings*
Indicator 1	0.52	0.35	-
Indicator 3	0.62	0.59	11.86
Indicator 4	0.63	0.65	11.68

6.2.4 Commitment

Following the conceptualization provided in chapter 4.2.4, commitment in the present study is limited to its affective component – normative and cognitive aspects of commitment are not captured. The operationalization of the commitment construct once again stems from the research conducted by Wallenburg (2004, pp. 189-190), who based his scale on German researcher Zimmer (2000). Investigating the role of commitment in business-to-business relationships, his affective commitment construct captures both a positive emotional attitude towards a supplier, as well as congruence between suppliers' and customers' goals. While Wallenburg embraced Zimmer's two items that refer to the customer's emotional attitude, he did not capture Zimmer's items on goal congruence, arguing that positive emotional attitudes and goal congruence are not necessarily present at the same time (Wallenburg 2004, p. 189).

Table 6-44. Indicators for the measurement of "commitment"

	Please indicate your level of agreement with the following statements regarding your logistics service provider.
Indicator 1	We come to our LSP's defense when it is criticized by persons from inside or outside our organization.
Indicator 2	We would be very sorry personally if we had to terminate the relationship with this LSP.
Indicator 3	We feel personally offended, when this LSP is criticized by persons from inside or outside of our company.
Indicator 4	We strongly intend to keep up the relationship with this LSP for as long as possible.

In addition to the two emotional attitude items (indicators 1 and 2 in Table 6-44), Wallenburg added an indicator originally introduced by Werner (1997, p. 148) to grasp the customer's motivation to maintain a relationship with a supplier even if this meets resistance (indicator 4). In the context of the present study, a fourth indicator is used which refers to an even stronger indication of commitment. Accordingly, indicator 3 measures the customers' emotional response yielded when the service provider is criticized.

Table 6-45. Adaptation of "commitment" (Germany, 4 indicators)

Information on the factor "commitment" (Germany, 4 indicators)			
Coefficient alpha	0.82	AGFI	0.95
Explained factor variance	55.72%	CFI	0.99
χ^2/df	5.282	RMSEA	0.089
TLI	0.97	Composite reliability	0.82
GFI	0.99	Average variance extracted	0.54
Information on the indicators			
	Item-total correlation	*Individual item reliability*	*t-value of factor loadings*
Indicator 1	0.66	0.52	-
Indicator 2	0.71	0.72	17.35
Indicator 3	0.53	0.31	12.06
Indicator 4	0.70	0.67	17.11

Table 6-46. Adaptation of "commitment" (USA, 4 indicators)

Information on the factor "commitment" (USA, 4 indicators)			
Coefficient alpha	0.82	AGFI	0.75
Explained factor variance	54.92%	CFI	0.94
χ^2/df	12.793	RMSEA	0.218
TLI	0.82	Composite reliability	0.83
GFI	0.95	Average variance extracted	0.56
Information on the indicators			
	Item-total correlation	*Individual item reliability*	*t-value of factor loadings*
Indicator 1	0.62	0.42	-
Indicator 2	0.68	0.66	10.03
Indicator 3	0.61	0.42	8.55
Indicator 4	0.70	0.69	10.11

As Table 6-45 and Table 6-46 show, the measurement of commitment in both Germany and the USA has some deficits. In both cases, indicator 3 exhibits a very low individual item reliability, paired with an item-total correlation that is lower than the ones of all other indicators. For this reason, the newly introduced indicator is eliminated, as it is likely that the emotional response of feeling offended can be considered too strong in the context of buyer-seller relationships.

Table 6-47. Adaptation of "commitment" (Germany, 3 indicators)

Information on the factor "commitment" (Germany, 3 indicators)			
Coefficient alpha	0.84	AGFI	*
Explained factor variance	63.65%	CFI	*
χ^2/df	*	RMSEA	*
TLI	*	Composite reliability	0.84
GFI	*	Average variance extracted	0.65
* Value cannot be calculated, because factor with 3 indicators does not possess degrees of freedom.			
Information on the indicators			
	Item-total correlation	*Individual item reliability*	*t-value of factor loadings*
Indicator 1	0.64	0.50	-
Indicator 2	0.74	0.74	16.38
Indicator 4	0.72	0.67	16.39

Table 6-48. Adaptation of "commitment" (USA, 3 indicators)

Information on the factor "commitment" (USA, 3 indicators)			
Coefficient alpha	0.80	AGFI	*
Explained factor variance	58.94%	CFI	*
χ^2/df	*	RMSEA	*
TLI	*	Composite reliability	0.82
GFI	*	Average variance extracted	0.62
* Value cannot be calculated, because factor with 3 indicators does not possess degrees of freedom.			
Information on the indicators			
	Item-total correlation	*Individual item reliability*	*t-value of factor loadings*
Indicator 1	0.54	0.36	-
Indicator 2	0.69	0.64	9.00
Indicator 4	0.73	0.77	8.71

While the resulting measurement with three indicators displays satisfactory fit in Germany, indicator 1 still causes some disturbance in the US measurement, as expressed through individual item reliability. In order not to limit content validity and because all other adaptation measures indicate satisfactory values, the model is not exposed to further modification.

6.2.5 Trust

In the understanding coined by Ganesan (1994), benevolence and credibility are both antecedents and consequences of trust. Consequently, both aspects have to be included in the operationalization of the construct. While prior research is not conclusive as to whether trust is two dimensional or not (Matthes 2006, p. 103), i.e. whether benevolence and credibility form two discriminant dimensions of trust, Wallenburg (2004, pp. 185-189) found evidence that they are not separable. He thus supports the findings of Doney and Cannon (1997), who regard trust to be uni-dimensional.

As it was stated in chapter 4.2.5, trustees can be both organizations and individuals. For both constructs, the operationalizations are based on those provided by Wallenburg (2004), which are based on scales developed by Doney and Cannon (1997, pp. 48-49) and complemented by indicators suggested by Werner (1997).

6.2.5.1 Organizational trust

Doney and Cannon (1997) developed a scale for trust in supplier firms which serves as the basis of Wallenburg's measurement instrument. In fact, of the indicators displayed in Table 6-49, indicators 1, 2, and 4 originate from Doney and Cannon (1997), while indicator 2 stems from Werner (1997). Indicator 5 was introduced by Wallenburg (2004) to complement the other indicators.

Table 6-49. Indicators for the measurement of "organizational trust"

	Please indicate your level of agreement with the following statements regarding your logistics service provider.
Indicator 1	This LSP keeps promises it makes to our firm.
Indicator 2	Regarding problems, this LSP is always honest with us.
Indicator 3	This LSP is trustworthy.
Indicator 4	This LSP its genuinely concerned that our business succeeds.
Indicator 5	The LSP correctly carries out tasks that we cannot directly control.

While indicators 1, 2, and 3 measure basic credibility, indicator 5 captures an additional perspective of credibility, as it refers to the trust that even tasks outside the direct control of the customer, which are in most instances costly to monitor, are carried out in the customer's interest. Indicator 4, on the other hand, captures benevolence, as it reflects the customer's trust that the supplier will actually have its benefits and goals in mind.

Table 6-50. Adaptation of "organizational trust" (Germany, 5 indicators)

Information on the factor "organizational trust" (Germany, 5 indicators)			
Coefficient alpha	0.91	AGFI	0.92
Explained factor variance	66.87%	CFI	0.98
χ^2/df	7.531	RMSEA	0.110
TLI	0.97	Composite reliability	0.91
GFI	0.97	Average variance extracted	0.66
Information on the indicators			
	Item-total correlation	*Individual item reliability*	*t-value of factor loadings*
Indicator 1	0.72	0.59	-
Indicator 2	0.83	0.83	23.41
Indicator 3	0.86	0.87	23.97
Indicator 4	0.70	0.53	17.85
Indicator 5	0.71	0.52	17.66

Table 6-51. Adaptation of "organizational trust" (USA, 5 indicators)

Information on the factor "organizational trust" (USA, 5 indicators)			
Coefficient alpha	0.92	AGFI	0.92
Explained factor variance	70.79%	CFI	0.99
χ^2/df	3.547	RMSEA	0.101
TLI	0.98	Composite reliability	0.92
GFI	0.97	Average variance extracted	0.70
Information on the indicators			
	Item-total correlation	*Individual item reliability*	*t-value of factor loadings*
Indicator 1	0.81	0.72	-
Indicator 2	0.86	0.87	20.55
Indicator 3	0.89	0.91	21.31
Indicator 4	0.78	0.65	15.85
Indicator 5	0.62	0.39	10.91

Table 6-50 shows that the measurement model in Germany possesses satisfactory adaptation in most dimensions – only adjusted χ^2 and RMSEA indicate some problems. This is attributable to a high correlation of the error term of indicator 4 with those of indicators 2 and 5, wherefore indicator 4 is eliminated from the model. The remaining four item measurement (Table 6-52) displays excellent model fit and does not require further modification.

In the USA (Table 6-51), similarly, adjusted χ^2 and RMSEA indicate problems, which become manifest in indicator 5's relatively low individual item reliability. For this reason, indicator 5 is eliminated from the measurement model in the USA and the remaining four item measurement is satisfactory (Table 6-53) – while adjusted χ^2 and RMSEA are still slightly above the recommended threshold values, further modification is not pursued in order not to impair content validity.

Table 6-52. Adaptation of "organizational trust" (Germany, 4 indicators)

Information on the factor "organizational trust" (Germany, 4 indicators)			
Coefficient alpha	0.90	AGFI	0.99
Explained factor variance	70.19%	CFI	1.00
χ^2/df	0.914	RMSEA	0.000
TLI	1.00	Composite reliability	0.90
GFI	1.00	Average variance extracted	0.70
Information on the indicators			
	Item-total correlation	*Individual item reliability*	*t-value of factor loadings*
Indicator 1	0.74	0.60	-
Indicator 2	0.85	0.86	23.90
Indicator 3	0.84	0.84	23.67
Indicator 5	0.68	0.51	17.45

Table 6-53. Adaptation of "organizational trust" (USA, 4 indicators)

Information on the factor "organizational trust" (USA, 4 indicators)			
Coefficient alpha	0.93	AGFI	0.93
Explained factor variance	78.32%	CFI	1.00
χ^2/df	3.372	RMSEA	0.098
TLI	0.98	Composite reliability	0.93
GFI	0.99	Average variance extracted	0.78

Information on the indicators			
	Item-total correlation	*Individual item reliability*	*t-value of factor loadings*
Indicator 1	0.81	0.71	-
Indicator 2	0.89	0.87	20.42
Indicator 3	0.91	0.91	21.14
Indicator 4	0.77	0.64	15.62

6.2.5.2 Personal trust

Wallenburg (2004, pp. 188-189) operationalizes personal trust following the operationalization of organizational trust, but recurring to the respondent's specific contact person at the LSP. For the present study (Table 6-54), indicators 2, 4, and 5 are taken from Wallenburg's scale, while indicator 1 was added on the basis of Doney and Cannon (1997, p. 49). In addition to grasping aspects of benevolence and credibility, the present scale also captures respondents' evaluations of contact persons' capabilities, as stated in indicator 3.

Table 6-54. Indicators for the measurement of "personal trust"

	Please indicate your level of agreement with the following statements regarding your contact person at the logistics service provider.
Indicator 1	This person is trustworthy.
Indicator 2	This person does not make false promises.
Indicator 3	This person is a highly knowledgeable manager.
Indicator 4	This person speaks the same "language" we do.
Indicator 5	We trust this person to keep our best interests in mind.

As it was the case with organizational trust, EFA confirms the construct's one dimensional structure. However, as displayed in Table 6-55 for Germany and in Table 6-56 for the USA, the measurement of personal trust is marked by some deficiencies that are stated in adjusted χ^2's and RMSEA's that exceed the recommended thresholds. In both countries, this can be ascribed to correlations of indicator 4's error term with another error term – in Germany with the error term of indicator 5 and in the USA with the error term of indicator 3.

Table 6-55. Adaptation of "personal trust" (Germany, 5 indicators)

Information on the factor "personal trust" (Germany, 5 indicators)			
Coefficient alpha	0.93	AGFI	0.93
Explained factor variance	73.71%	CFI	0.99
χ^2/df	6.968	RMSEA	0.105
TLI	0.97	Composite reliability	0.93
GFI	0.98	Average variance extracted	0.74
Information on the indicators			
	Item-total correlation	*Individual item reliability*	*t-value of factor loadings*
Indicator 1	0.84	0.77	-
Indicator 2	0.83	0.75	27.67
Indicator 3	0.80	0.69	25.82
Indicator 4	0.82	0.73	27.05
Indicator 5	0.82	0.74	27.42

Table 6-56. Adaptation of "personal trust" (USA, 5 indicators)

Information on the factor "personal trust" (USA, 5 indicators)			
Coefficient alpha	0.94	AGFI	0.85
Explained factor variance	76.79%	CFI	0.98
χ^2/df	6.283	RMSEA	0.146
TLI	0.95	Composite reliability	0.94
GFI	0.95	Average variance extracted	0.77
Information on the indicators			
	Item-total correlation	*Individual item reliability*	*t-value of factor loadings*
Indicator 1	0.87	0.81	-
Indicator 2	0.84	0.76	20.13
Indicator 3	0.87	0.81	21.60
Indicator 4	0.83	0.75	19.80
Indicator 5	0.82	0.72	19.11

Table 6-57. Adaptation of "personal trust" (Germany, 4 indicators)

Information on the factor "personal trust" (Germany, 4 indicators)			
Coefficient alpha	0.92	AGFI	1.00
Explained factor variance	73.84%	CFI	1.00
χ^2/df	0.103	RMSEA	0.000
TLI	1.00	Composite reliability	0.92
GFI	1.00	Average variance extracted	0.74
Information on the indicators			
	Item-total correlation	*Individual item reliability*	*t-value of factor loadings*
Indicator 1	0.84	0.80	-
Indicator 2	0.83	0.77	28.50
Indicator 3	0.78	0.68	25.30
Indicator 5	0.80	0.70	26.19

Without limiting the models' content validities, indicator 4 is excluded for Germany and the USA alike. The resulting measurement models, consisting of indicators 1, 2, 3, and 5, and exhibited in Table 6-57 and in Table 6-58, display excellent model fit.

Table 6-58. Adaptation of "personal trust" (USA, 4 indicators)

Information on the factor "personal trust" (USA, 4 indicators)			
Coefficient alpha	0.93	AGFI	0.97
Explained factor variance	77.50%	CFI	1.00
χ^2/df	1.704	RMSEA	0.053
TLI	1.00	Composite reliability	0.93
GFI	0.99	Average variance extracted	0.77
Information on the indicators			
	Item-total correlation	*Individual item reliability*	*t-value of factor loadings*
Indicator 1	0.87	0.83	-
Indicator 2	0.84	0.78	21.09
Indicator 3	0.83	0.75	20.21
Indicator 5	0.82	0.73	19.56

6.2.6 Alternatives

As pointed out before, the existence of adequate alternatives to the presently used LSP is the prerequisite for churning – only if true alternatives exist can a customer be 'un-loyal' and switch to another service provider. Therefore, the measurement of alternatives necessitates the evaluation of the presence of the availability of LSPs that offer the same service at the same price. In this respect, a single item measurement is sufficient for capturing the intended content. As the market for logistics services is marked by a high degree of competition, it can be assumed that there always are alternate LSPs which can deliver a certain service. Wallenburg (2004, p. 201) points out that these LSPs available in the market form a continuum that ranges from unattractive LSPs, characterized by high prices and/or insufficient performance, to attractive LSPs, which offer very good prices and/or very high performance. For the purposes of the present study, it is necessary to determine the relative position of the currently used LSP within the continuum of market alternatives. This is achieved through the indicator displayed in Table 6-59.

Table 6-59. Indicator for the measurement of "alternatives"

	Please indicate your level of agreement with the following statements on your attitude towards this LSP and on market alternatives.
Indicator 1	There are many alternative LSPs that can offer the same service at the same cost.

Assessing model fit is not possible in single item measurements. However, the calculation of the average variance extracted is necessary for conducting analyses on discriminant validity using the Fornell/Larcker criterion. As stated previously (chapter 5.4.4.2), factor loadings, item measurement errors, and the construct's variance are needed to compute AVE. While an item's factor loading is 1 in single item measurements, its measurement error cannot be determined in the absence of further indicators. This problem is discussed by Wallenburg (2004, p. 161), who explains that measurement errors are a result of textual and formal deficiencies. The former are usually the result of textual differences between the indicators forming a construct that prevent an exact match between indicators and the underlying construct. While this phenomenon is ruled out in single item measurements, where textually the construct is per definition equal to the single item, formal measurement error may still be present. Therefore, the approach proposed by Wallenburg (2004, p. 161) is followed, who maintains that the value of the measurement error in single item measurements should be estimated by examining the best indicators of the other constructs. As in his study, the best indicators here have measurement errors of approximately 0.2, which will thus be used as an estimate in the following analyses.[3]

6.2.7 Discriminant validity

A final step in the operationalization of measurement models is the examination of discriminant validity, which is performed by applying the Fornell/Larcker criterion.

In the German sample, fit criteria of the CFA conducted on the 9-factor model show that all requirements are met, except for GFI and AGFI, which remain slightly below the recommended threshold value of 0.9 (see Table 6-60). Due to the model's complexity, however, this is tolerable, and a first indication of discriminant validity is therefore identified.

As a second step, squared correlations between constructs are compared with the corresponding AVEs. This analysis reveals that problems with discriminant validity exist with regard to the two constructs of fairness and organizational trust (Table 6-61). Fairness appears to be indiscriminant from repurchase intentions, relational satisfaction, service quality, personal trust, and organizational trust. In fact, EFA reveals that significant cross loadings of fairness indicators to the other constructs exist. Especially with regard to relational satisfaction, EFA indicates higher loadings of fairness

[3] Conducting a CFA with indicator 1's measurement error set to 0.2 yields an AVE of 0.922. This value will be used subsequently to assess discriminant validity using the Fornell/Larcker criterion.

indicators on relational satisfaction than on fairness itself. Also, fairness possesses the smallest AVE of all measured constructs, indicating that the construct is not too well represented by its indicators.

Table 6-60. Fit of the 9-factor CFA (Germany)

Fit of the 9-factor customer loyalty model (Germany)			
Coefficient alpha	*	AGFI	0.87
Explained factor variance	*	CFI	0.97
χ^2/df	2.053	RMSEA	0.044
TLI	0.96	Composite reliability	*
GFI	0.90	Average variance extracted	*
* This value can only be calculated on the factor level.			

Table 6-61. Test for discriminant validity (Germany)

		Repurchase	Additional Purchase	Referrals	Price Satisfaction	Relational Satisfaction	Service Quality	Proactive Improvement	Fairness	Commitment	Personal Trust	Organizational Trust
	AVE	0.80	0.67	0.85	0.70	0.76	0.86	0.72	0.51	0.65	0.74	0.70
Repurchase	0.80											
Additional Purchase	0.67	0.22										
Referrals	0.85	0.36	0.29									
Price Satisfaction	0.70	0.33	0.11	0.21								
Relational Satisfaction	0.76	0.59	0.12	0.27	0.31							
Service Quality	0.86	0.57	0.17	0.36	0.32	0.64						
Proactive Improvement	0.72	0.35	0.21	0.34	0.16	0.30	0.44					
Fairness	0.51	**0.61**	0.11	0.26	0.38	**0.82**	**0.56**	0.33				
Commitment	0.65	0.58	0.23	0.43	0.29	0.54	0.48	0.26	0.50			
Personal Trust	0.74	0.59	0.15	0.31	0.34	0.63	0.64	0.37	**0.61**	0.55		
Organizational Trust	0.70	0.53	0.13	0.27	0.29	**0.78**	0.65	0.34	**0.73**	0.55	**0.73**	
Alternatives	0.92	0.15	0.04	0.03	0.11	0.08	0.05	0.02	0.10	0.09	0.06	0.05
Bold values indicate squared correlations that exceed the corresponding AVEs												

While there still is substantiative reasoning for expecting fairness to be an important factor for explaining customer loyalty, the chosen operationalization appears to allow for too much similarity with other factors. For this reason, the construct cannot be included in the structural analyses of the following sections. For future research, however, the operationalization of fairness should be reconsidered and indicators should be selected that provide a higher degree of uniqueness.

The second construct implying shortcomings of discriminance is organizational trust. In addition to being non-discriminant from fairness, as pointed out above, problems arise with personal trust and relational satisfaction. Although textually, organizational trust is clearly different from personal trust, as the trustee of the former is the organization of the LSP has a whole, while it is the specific contact person for the latter, respondents do not appear to distinguish this clearly. While the high correlation may be a result of a strong influence of personal trust on organizational trust, as hypothesized earlier, organizational trust is also similar to relational satisfaction. Even though the hypothesized model of customer loyalty assumes relational satisfaction to positively influence trust, the current operationalization of the organizational trust construct evidently is not sufficiently different from relational satisfaction. As a consequence, organizational trust is excluded from further analyses. Nevertheless, the textual consequences of the exclusion are not serious, as the notion of trust is still represented by personal trust, which was anyway assumed to be an antecedent of organizational trust.

Table 6-62. Fit of the 9-factor CFA (USA)

Fit of the 9-factor customer loyalty model (USA)			
Coefficient alpha	*	AGFI	0.82
Explained factor variance	*	CFI	0.95
χ^2/df	1.586	RMSEA	0.049
TLI	0.96	Composite reliability	*
GFI	0.85	Average variance extracted	*
* This value can only be calculated on the factor level.			

For the USA, fit criteria of the CFA are similar to those in Germany and all indices except GFI and AGFI fulfill the requirements (Table 6-62). By examining AVEs and squared correlations in the USA, however, problems with fairness and commitment become visible (Table 6-63). While the same reasoning for eliminating fairness as provided above has to be applied to the US sample, commitment shows to be non-discriminant from relational satisfaction, service quality, and organizational trust.

The lacking discriminance of commitment from these constructs is mainly due to the construct's relatively low AVE, which is, in fact, the lowest AVE observed in the USA. Hence, the indicators constituting commitment do not appear to provide a good representation of the construct and some commitment indicators, most prominently indicator 2, show strong cross loadings onto other factors, leading to the insufficient discriminance. As a result, commitment, although based on a sound conceptualization and operationalization, has to be excluded from further analyses on the US sample. Interestingly, this deficiency does not surface in the German sample, which may be a result of the operationalization of the construct that is rooted in German research on commitment.

Table 6-63. Test for discriminant validity (USA)

		Loyalty	Price Satisfaction	Relational Satisfaction	Service Quality	Proactive Improvement	Fairness	Commitment	Personal Trust	Organizational Trust
	AVE	0.65	0.73	0.78	0.88	0.74	0.55	0.51	0.77	0.79
Loyalty	0.65									
Price Satisfaction	0.73	0.43								
Relational Satisfaction	0.78	0.55	0.47							
Service Quality	0.88	**0.68**	0.54	0.62						
Proactive Improvement	0.74	0.41	0.32	0.39	0.46					
Fairness	0.55	0.50	0.51	**0.80**	0.54	0.43				
Commitment	0.51	0.51	0.40	**0.63**	**0.54**	0.35	**0.58**			
Personal Trust	0.77	0.39	0.39	0.49	0.39	0.31	0.47	0.43		
Organizational Trust	0.79	0.45	0.41	0.74	0.48	0.29	**0.73**	**0.52**	0.54	
Alternatives	0.92	0.06	0.05	0.05	0.04	0.03	0.08	0.06	0.04	0.06
Bold values indicate squared correlations that exceed the corresponding AVEs										

Additionally, partial evidence of non-discriminance of service quality and loyalty is obtained in the USA. However, these two constructs are textually different and EFA does not provide indication of substantial cross loadings between the two constructs. Consequently, action is not necessary and both constructs will be used in further analyses.

6.3 Moderators

This section will operationalize the four moderators examined in this study. While opportunism is measured as a latent variable, tending to the textual complexity of the issue, the simplicity of the remaining three moderators does not necessitate multi-item measurement models.

6.3.1 Opportunism

Opportunism is measured through a synthesis of the scales proposed by Morgan and Hunt (1994) and Knemeyer and Murphy (2004), such that the central aspects of opportunistic behavior are captured. While the three-item scale of Morgan and Hunt (1994, p.35) is not broad enough to capture all aspects conceptualized in chapter 4.4.1.1, the nine-item scale used by Knemeyer and Murphy (2004, p.51) is overly complex. Therefore, the essential elements of both scales are merged into the scale depicted in Table 6-64.

Table 6-64. Indicators for the measurement of opportunism

	Please indicate your level of agreement with the following statements on your relationship with this logistics service provider.
Indicator 1	To accomplish its own objectives, sometimes our LSP alters the facts slightly.
Indicator 2	To accomplish its own objectives, sometimes our LSP promises to do things without actually doing them later.
Indicator 3	Our LSP sometimes exaggerates its requirements in order to get what it really needs from us.
Indicator 4	Our LSP feels that it is alright to do anything within its means to further its own interests.
Indicator 5	Our LSP feels that honesty does pay when dealing with us.

Morgan and Hunt (1994) and Knemeyer and Murphy (2004) agree that the alteration of facts by one party with the aim of reaching its own objectives is an important characteristic of opportunistic behavior. This is reflected by indicators 1 and 2, adapted only to provide a better language fit with the other indicators. The remaining three indicators are all taken from the scale proposed by Knemeyer and Murphy (2004, p. 51). Indicator 3 measures the truthfulness and accuracy of the LSP's portrayal of its activities. If the LSP is deceiving the customer by not providing correct information on its activities, the customer will be discriminated, because it can neither reliably assess logistics performance, nor logistics costs. An even stronger perception of dishonesty is grasped by indicator 4 and finally, indicator 5 (reverse coded) measures the degree to which the customer feels that its LSP ascribes high importance to honest business attitude. If this perception is positive, the risk of opportunistic behavior can be surmised to be substantially decreased. As such, the entire scale grasps customers' perceptions on LSPs' truthfulness and honesty, as well as the reliability and accurateness of information provided and promises made by LSPs.

As reported in Table 6-65 for Germany and in Table 6-66 for the USA, both measurement models fail to fulfill the fit-requirements. Measurement in both cases is marked by very low individual item reliabilities of the reverse-coded indicator 5. While explained factor variance in Germany is below the 50% threshold, leading to the elimination of indicator 5 there, the item's reliability is extremely low at a value of 0.05 in the USA, wherefore it is also excluded from analyses in the USA.

Table 6-65. Adaptation of "opportunism" (Germany, 5 indicators)

Information on the factor "opportunism" (Germany, 5 indicators)			
Coefficient alpha	0.82	AGFI	0.97
Explained factor variance	48.83%	CFI	0.99
χ^2/df	2.247	RMSEA	0.048
TLI	0.99	Composite reliability	0.83
GFI	0.99	Average variance extracted	0.51
Information on the indicators			
	Item-total correlation	*Individual item reliability*	*t-value of factor loadings*
Indicator 1	0.60	0.45	-
Indicator 2	0.70	0.63	15.17
Indicator 3	0.72	0.68	15.44
Indicator 4	0.57	0.41	12.83
Indicator 5	0.47	0.27	10.72

Table 6-66. Adaptation of "opportunism" (USA, 5 indicators)

Information on the factor "opportunism" (USA, 5 indicators)			
Coefficient alpha	0.81	AGFI	0.95
Explained factor variance	55.44%	CFI	0.99
χ^2/df	2.325	RMSEA	0.073
TLI	0.98	Composite reliability	0.83
GFI	0.98	Average variance extracted	0.52
Information on the indicators			
	Item-total correlation	*Individual item reliability*	*t-value of factor loadings*
Indicator 1	0.73	0.71	-
Indicator 2	0.74	0.71	15.67
Indicator 3	0.72	0.71	15.59
Indicator 4	0.70	0.59	13.76
Indicator 5	0.21	0.05	3.31

Table 6-67. Adaptation of "opportunism" (Germany, 4 indicators)

Information on the factor "opportunism" (Germany, 4 indicators)			
Coefficient alpha	0.82	AGFI	0.99
Explained factor variance	54.22%	CFI	1.00
χ^2/df	1.662	RMSEA	0.035
TLI	1.00	Composite reliability	0.82
GFI	1.00	Average variance extracted	0.54
Information on the indicators			
	Item-total correlation	*Individual item reliability*	*t-value of factor loadings*
Indicator 1	0.60	0.44	-
Indicator 2	0.68	0.60	14.65
Indicator 3	0.72	0.70	15.10
Indicator 4	0.58	0.42	12.77

Table 6-67 and Table 6-68 report on the adaptation of the 4-item measurement in Germany and the USA. While the German model exhibits very good fit, the elimination of indicator 5 in the USA raised some adaptation measures over the recommended threshold values. Since adjusted χ^2 and RMSEA only slightly exceed the recommendations, though, and because all other fit indices demonstrate excellent fit, no further modification is needed.

Table 6-68. Adaptation of "opportunism" (USA, 4 indicators)

Information on the factor "opportunism" (USA, 4 indicators)			
Coefficient alpha	0.89	AGFI	0.93
Explained factor variance	68.08%	CFI	0.99
χ^2/df	3.837	RMSEA	0.107
TLI	0.97	Composite reliability	0.90
GFI	0.99	Average variance extracted	0.68
Information on the indicators			
	Item-total correlation	*Individual item reliability*	*t-value of factor loadings*
Indicator 1	0.78	0.71	-
Indicator 2	0.78	0.71	15.63
Indicator 3	0.79	0.71	15.62
Indicator 4	0.72	0.59	13.71

6.3.2 Relationship age and centralization

Relationship age refers to the duration of the working relationship between a customer and its most important LSP, for which the respondent provided information in the survey. Due to the simplicity of the issue, a multi-item scale was not employed and instead, an open question was asked, in which respondents were asked to indicate the number of years they have been working together with their most important LSP – one decimal was allowed to grasp fractions of a year. The respective question reads: "For how long have you been working together with this LSP?"

Similarly, the centralization of logistics decisions is captured using the following question:

Table 6-69. Measurement of the centralization of logistics decisions

	Please indicate your level of agreement with the following statements about your internal logistics organization.
Indicator 1	The decision-making process on logistics issues is very centralized in our company.

Analogous to latent variables, respondents were asked to indicate their level of agreement with the proposed statement on a seven-point Likert-scale (1 = strongly disagree; 7 = strongly agree).

6.3.3 Outsourcing focus

The outsourcing focus is concerned with the area, in which a respondent's most important LSP renders services. To collect this information, respondents were presented with a list of 13 logistics activities that an LSP could potentially provide (adapted from Knemeyer, Corsi, and Murphy 2003). Then, respondents were asked to indicate the degree of execution by the most important LSP for each service (7-point Likert-scale, from 1 = not at all, to 7 = completely). In addition, respondents were requested to mark for each of the services, whether the service's focus was inbound, internal, or outbound. In this way, information was collected on the degree to which different services are carried out by the focal LSP and where the outsourcing focus is for each service.

Services	Degree of execution	Focus of service
	1=not at all 7=completely	1=inbound 2=internal 3=outbound
Service 1	3	1
Service 2	7	3
Service 3	1	2
Service 4	4	3
Service 5	5	1

Services	Inbound	Internal	Outbound
Service 1	3		
Service 2			7
Service 3		1	
Service 4			4
Service 5	5		
Sum	8	1	11

Fig. 6-1. Determination of the outsourcing focus (example)

In order to use this information as a moderator, the outsourcing focus information available for each service has to be condensed to an individual score for each case. Figure 6-1 describes the procedure performed to accomplish this. For each case (left side of Figure 6-1), Likert-scores from the service execution question are added up, distinguishing between inbound, internal, and outbound foci (right side of Figure 6-1). Then, a score is assigned for the overall outsourcing focus, representing the highest count from the performed aggregation, which would be 11 in the example, such that the outsourcing focus of that respondent is termed outbound. Another calculation method would be the combination of scores for inbound and outbound foci already at this point. Since the overall classification is nearly similar, with a difference of only 3 (4) cases in Germany (the USA), scores will only be aggregated later, such that information can be reported for the three outsourcing foci separately (see Figure 6-2).

Figure 6-2 demonstrates the outsourcing foci in Germany and the USA. In both countries, projects with the most important LSP that are outbound-focused clearly dominate logistics outsourcing. However, this focus is even more emphasized in Germany, where three quarters of respondent companies engage in outbound outsourcing. In Germany, internal tasks are rarely in the focus of firms' outsourcing – not even 10% of respondents indicated that their most important LSP was rendering services internally. In the USA, the share of internally focused projects is considerably higher at approximately 20%.

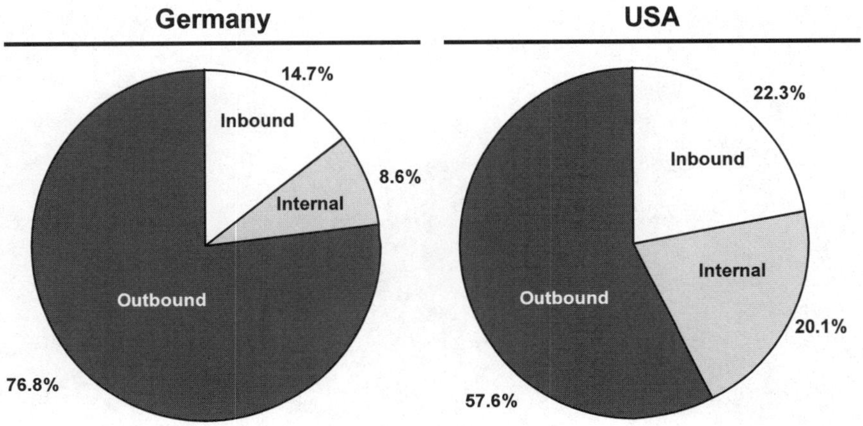

Fig. 6-2. Outsourcing focus in Germany and the USA

For the purpose of conducting moderation analyses to test the hypothesis stated in chapter 4.4.2.4, information on the outsourcing focus has to be further aggregated, as only internal and external foci, the latter consisting of both inbound and outbound foci, are compared. For this, a new score is introduced for each case, in which inbound and outbound foci are combined, and juxtaposed to internal arrangements.

7 Structural models

7.1 Results from Germany

7.1.1 Customer loyalty model in Germany

As a result of the testing for discriminance performed in chapter 6.2.7, the constructs of fairness and organizational trust had to be excluded from further analyses of the German sample. Consequently, seven determinants are included in the structural equation model employed to explain the three dimensions of customer loyalty, repurchase intentions, additional purchase intentions, and referral behavior. The underlying causal structures, after elimination of fairness and organizational trust, are shown in Figure 7-1 (in addition to the indicated paths, all factors are hypothesized to directly influence each of the three customer loyalty dimensions).

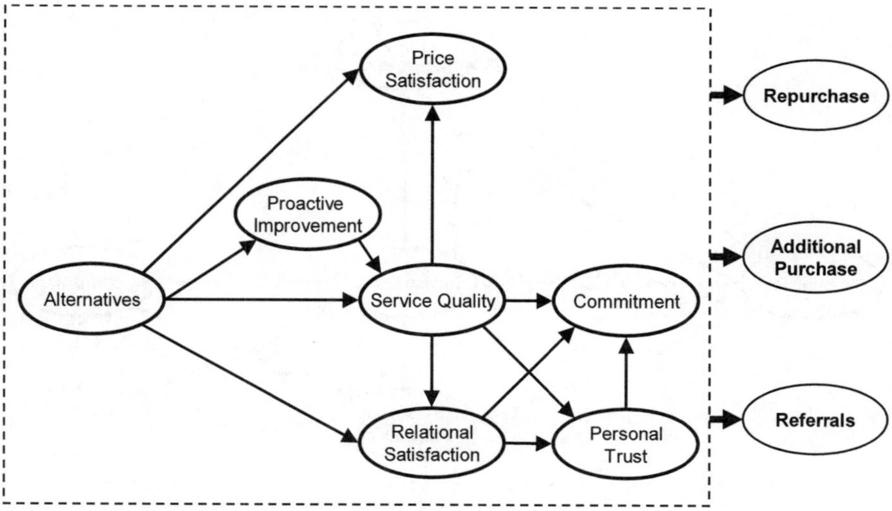

Fig. 7-1. Revised hypothesized model of customer loyalty (Germany)

Applying structural equation modeling to the hypothesized model structure yields a very good model fit, as shown in Table 7-1. Adaptation measures demonstrate that the model fits the empirical data in Germany very well and even though AGFI is slightly below the recommended threshold value, model refitting is not required.

Table 7-1. Adaptation of the 7-factor loyalty model (Germany)

Adaptation measures						Explanatory power (R^2)		
χ^2/df	TLI	GFI	AGFI	CFI	RMSEA	Repurchase	Additonal Purchase	Referrals
2.212	0.964	0.913	0.890	0.970	0.047	74.3%	32.9%	54.3%

Examining squared multiple correlations for the three customer loyalty dimensions reveals that very good explanatory power is provided for repurchase intentions, while additional purchase intentions and referral behavior are only partially explained. In the following sections, results contained in the structural equation model will be explained. While chapter 7.1.1.1 discusses findings on relationships within the model, chapters 7.1.1.2 to 7.1.1.4 are dedicated to an examination of the effects on customer loyalty.

7.1.1.1 Relationships between determinants

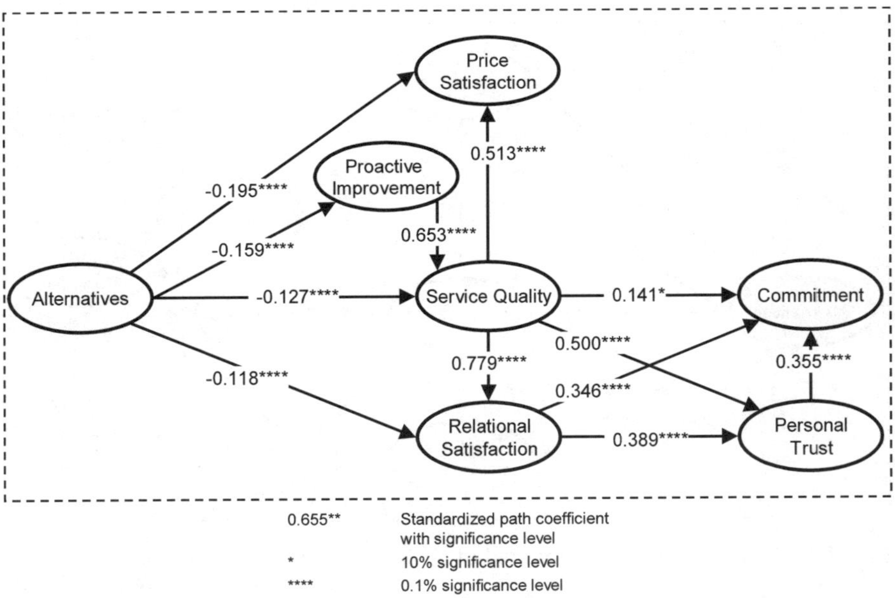

Fig. 7-2. Relationships between loyalty determinants (Germany)

The model of customer loyalty, as displayed in Figure 7-2, contains 12 linkages between the seven determinants of customer loyalty. All of these are significant – the four effects of alternatives on proactive improvement, price satisfaction, service quality, and relational satisfaction are negative, while the remaining eight linkages are positive. Thus, the underlying 12 hypotheses are supported by the data gathered in Germany, which is stated in Table 7-2.

Table 7-2. Hypothesis-tests for linkages between determinants (Germany)

Effects \ Effects on	Service Quality		Price Satisfaction		Relational Satisfaction		Proactive Improvement		Fairness		Commitment		Personal Trust		Organizational Trust	
Service Quality			H_{D4}	✓	H_{D5}	✓			H_{D6}	×	H_{D2}	✓	H_{D3a}	✓	H_{D3b}	×
Price Satisfaction									H_{D8}	×						
Relational Satisfaction									H_{D12}	×	H_{D10}	✓	H_{D11a}	✓	H_{D11b}	×
Proactive Improvement	H_{D14}	✓							H_{D15}	×						
Fairness											H_{D17}	×	H_{D18a}	×	H_{D18b}	×
Commitment																
Personal Trust											H_{D23}	✓			H_{D22}	×
Organizational Trust											H_{D24}	×				
Alternatives	H_{D27}	✓	H_{D28}	✓	H_{D29}	✓	H_{D26}	✓								

✓ Hypothesis cannot be rejected
⊘ Hypothesis is rejected
× Hypothesis cannot be tested

As suggested by social exchange theory, and reflected in Hypotheses H_{D26}, H_{D27}, H_{D28}, and H_{D29}, the availability of adequate alternatives influences the evaluation of satisfaction and proactive improvement. When assessing the current service provider, it is not sufficient for a customer to compare its realized benefit (B) to a comparison level (CL), which stands for the benefit potentially extractable from the relationship. Instead, B is compared to the best possible alternative, expressed through the comparison level of alternatives (CL_{alt}). This process is reflected in the data collected in Germany, which shows that the existence of adequate alternatives negatively impacts satisfaction with the relationship. Thus, a competitive market environment leads to increasing expectations from customers and, in consequence, LSPs in such an environment have to invest more into existing relationships than LSPs in markets, where adequate alternatives are not available.

It was also suspected that proactive improvement has two effects – on the one hand, proactive improvement efforts increase service quality, as an LSP independently pursues the improvement of its customer's processes. On the other hand, proactive improvement is expected to cause a positive signaling effect, by showing a customer that its LSP is truly concerned with creating benefits for the customer. While the former effect, the resulting increase of service quality, is captured through H_{D14} in the model, the latter effect is harder to grasp and reflected in the direct linkages between proactive improvement and customer loyalty. While the signaling effect will be discussed later on in the sections on loyalty effects, it is already clear now that proactive improvement is the decisive driver of service quality. With a standardized path coefficient of 0.653 at the 0.1% significance level, one of the highest values in the model is observed for this path.

Hypotheses H_{D4} and H_{D5} refer to the effects of service quality on price satisfaction and relational satisfaction, respectively. For both of these relationships, very strong support is offered by the data. As expected, service quality strongly influences the perception of price satisfaction, as customers are ultimately satisfied with the price offered by their LSP if they feel that the price they pay is in line with the service quality they receive. Also, good service quality drives relational satisfaction – a result of fewer conflicts arising from quality problems and a general cognition that the relationship yields the service quality that the customer feels entitled to.

Personal trust is created through the provision of good service quality (H_{D3a}) and the existence of relational satisfaction (H_{D11a}). While both effects are strong and highly significant, the empirical results indicate that the former relationship is much stronger than the latter, which is consistent with the previously explained finding that service quality enhances relational satisfaction. In addition to directly affecting personal trust, service quality has an influence on trust that is mediated through relational satisfaction and the total standardized effect of service quality on personal trust (0.803) is considerably higher than service quality's direct effect of 0.500.

Finally, evidence is found for the effects of service quality (H_{D2}), relational satisfaction (H_{D10}), and personal trust (H_{D23}) on (affective) commitment. While the direct effect of service quality on commitment is relatively week with a standardized path coefficient of 0.141 and a significance level of only 10%, an examination of total effects denotes a mediation of the effect of service quality on commitment through relational satisfaction and personal trust, expressed through a standardized total effect of 0.696. In addition, strong support is found for the postulates of commitment-trust theory, i.e. the existence of a strong influence of personal trust on the evaluation of commitment.

Comparing these results to those obtained by Wallenburg (2004, pp. 234-236), two major structural differences are found. First, the present model contains 'Personal Trust', which was not identified to add explanatory value to Wallenburg's model, leading to its elimination there. This can probably be attributed to the larger sample size of this study, which may cause more paths to be significant. Second, Wallenburg's final model contained the construct of 'Fairness', which had to be eliminated in this study due to its lacking discriminance from other constructs.

Comparing the individual paths shows that both models are very similar, with only a few differences in the paths themselves or the strengths of the linkages. With regard to alternatives, the present model contains four effects, which were all also part of Wallenburg's model. However, Wallenburg was not able to show a significant path between alternatives and relational satisfaction, which is highly significant in the present model. On the other hand, Wallenburg identified a relatively strong effect of alternatives on proactive improvement (-0.33****), which is significantly lower here (-0.159****). The remaining two effects of alternatives, finally, are more significant in the present than in Wallenburg's study, potentially attributable to the larger sample size of this study.

A nearly perfect consistency is found for the effects of proactive improvement on service quality and of service quality on relational satisfaction, which are strong and highly significant in both models. Deviating from Wallenburg's hypotheses, the present study proposes a link between service quality and price satisfaction, for which strong empirical support is found in the form of a path coefficient of 0.513, significant at the 0.1%-level.

Differences between Wallenburg's model and the present model are also apparent with respect to the effects of service quality and relational satisfaction on commitment. In both cases, effects in Wallenburg's model are considerably stronger than in the present model. This difference, however, can be attributed to the inclusion of personal trust in the present model, which mediates some of the effects of service quality and relational satisfaction on commitment.

As shown above, the formulated interdependencies among loyalty determinants are both supported by the empirical data and largely consistent with Wallenburg's findings. The following sections will therefore discuss the effects of the seven loyalty determinants on the three dimensions of customer loyalty and compare findings to those obtained by Wallenburg (2004).

7.1.1.2 Effects on repurchases

With regard to repurchases, 74.3% of the construct's variance is explained by the seven determinants of customer loyalty contained in the model, which means that only 25.7% are attributable to other factors. This is slightly lower than the R^2 of 83% yielded by Wallenburg's (2004, p. 237) model. Examining the individual linkages of the seven determinants and repurchase intentions (as reported in Table 7-3) reveals that all respective hypotheses (H_{D1a}, H_{D7a}, H_{D9a}, H_{D13a}, H_{D19a}, H_{D20a}, H_{D25a}) find support in the present study.

The three dimensions of satisfaction, service quality (H_{D1a}), price satisfaction (H_{D7a}), and relational satisfaction (H_{D9a}), positively influence repurchase intentions. It is interesting to note, however, that price satisfaction has a very weak effect, which is significant only at the 10% level. This is although most companies indicated that cost reductions were their most important reasons for engaging in logistics outsourcing (see Figure 2-4). The data, on the contrary, suggests that while cost reductions and prices play an important role in the initial outsourcing decision, they do not nearly as strongly motivate customers to maintain their relationships with LSPs.

Table 7-3. Effects of determinants on customer loyalty (Germany)

	Repurchases		Additional Purchases		Referrals	
	Direct Effect	Total Effect	Direct Effect	Total Effect	Direct Effect	Total Effect
Service Quality	0.16**	0.65	n.s.	0.18	0.19**	0.41
Price Satisfaction	0.05*	0.05	n.s.	0.06	0.09***	0.09
Relational Satisfaction	0.15**	0.37	-0.22**	-0.02	-0.17**	0.05
Proactive Improvement	0.11***	0.53	0.32****	0.44	0.30****	0.57
Commitment	0.31****	0.31	0.48****	0.48	0.54****	0.54
Personal Trust	0.17***	0.28	n.s.	0.10	n.s.	0.10
Alternatives	-0.07****	-0.37	n.s.	-0.17	n.s.	-0.15

0.16** Standardized path coefficient with significance level

0.65 Standardized total effect

n.s.;*/**/***/**** not significant; significant at the 10%/ 5%/ 1%/ 0.1% level

Proactive improvement was conceptualized to have two effects – first, to increase service quality and second, to send a positive signal to the customer that the LSP is striving to excel in the customer's best interest. While the performance enhancing effect was already identified in the previous section, the signaling effect is confirmed through proactive improvement's direct – and highly significant – influence on repurchase intentions (H_{D13a}).

The effect of commitment on repurchase intentions (H_{D19a}) is particularly strong and significant. This supports the expectation that customers maintain a relationship with a service provider because they affectively want to do so. Similarly, personal trust has a decisive effect on repurchase intentions (H_{D20a}). While Morgan and Hunt (1994) only found indication of an indirect effect on retention, mediated through commitment, a trustful relationship in itself clearly fosters relationship maintenance.

Finally, the availability of adequate alternatives to the current LSP has the hypothesized effect on repurchase intentions (H_{D25a}), i.e. the existence of such alternatives negatively affects repurchase intentions. While the direct effect is small, it is highly significant at the 0.1% level, indicating that alternatives indeed factor into customers' maintenance considerations.

Table 7-4 provides an overview of the hypothesis testing conducted for the direct effects on repurchase intentions:

Table 7-4. Hypothesis-tests for effects on repurchases (Germany)

Effects of \ Effects on	Repurchases	
Service Quality	H_{D1a}	✓
Price Satisfaction	H_{D7a}	✓
Relational Satisfaction	H_{D9a}	✓
Proactive Improvement	H_{D13a}	✓
Fairness	H_{D16a}	×
Commitment	H_{D19a}	✓
Personal Trust	H_{D20a}	✓
Organizational Trust	H_{D21a}	×
Alternatives	H_{D25a}	✓

✓ Hypothesis cannot be rejected
⊘ Hypothesis is rejected
× Hypothesis cannot be tested

As expected, total effects of service quality, relational satisfaction, proactive improvement, personal trust, and alternatives on repurchase intentions are considerably stronger than the direct effects, meaning that a larger part of these factors' influence is mediated through the other factors in the model. Only the total effects of price satisfaction and commitment on repurchase intentions are not increased by mediation. In the case of price satisfaction, this further supports the conclusion that prices only play a minor role in customers' considerations once immersed in a relationship with an LSP. The effect of commitment, on the other hand, cannot be reinforced by mediation, because the direct effects on the three dimensions of customer loyalty are the only effects emanating from commitment. It is important to note, however, that service quality and proactive improvement exhibit by far the strongest total effects on repurchase intentions. With values of 0.65 and 0.53 respectively, these two factors have decisive influences on customers' evaluations of whether to maintain a logistics outsourcing relationship or not.

While the interdependencies between the customer loyalty dimensions of the present study are highly consistent with those of Wallenburg (2004), the effects on repurchase intentions show considerable differences. As pointed out above, all respective hypotheses for direct effects find support by the data of the current study. In contrast, Wallenburg (2004, p. 238) only found significant direct loyalty effects for price satisfaction, fairness, commitment, and alternatives – effects of service quality, relational satisfaction, and proactive improvement were not significant. In addition, those direct effects that Wallenburg was able to validate differ from the findings of the current study. With the exception of the effect of alternatives, which is consistent with Wallenburg's results, effects of price satisfaction and commitment on repurchase intentions are significantly stronger in Wallenburg's study. This is in line with the observed total effects, which are consistent for service quality, relational satisfaction, and alternatives, while they are stronger in Wallenburg's model for price satisfaction and commitment, and weaker for proactive improvement. Again, this may in part be caused by the inclusion of personal trust in the present model, which has relatively strong effects, both direct and total.

7.1.1.3 Effects on additional purchases

The seven determinants of customer loyalty only explain 32.9% of the variance of additional purchase intentions, which matches Wallenburg's (2004, p. 237) R^2 of 31.2%. Hence, the hypothesized model only offers partial explanation of this loyalty aspect. Examining the corresponding direct effects of the seven factors on additional purchase intentions provides support for this finding, as only two out of seven hypotheses are supported by the data.

While the model's applicability to additional purchase intentions is thus limited, the roots of the problem can be attributed to the nature of the decision to outsource additional logistics tasks. Regarding repurchases, customers have already decided to outsource the respective service and therefore only have to choose between three options: remaining with the current LSP, switching to an alternative LSP, or in-sourcing the task. Additional purchase considerations, however, are more complex, as customers are faced with two different options: (a) A service is already outsourced, but currently provided by another LSP, i.e. an LSP that is not the focal LSP selected for the survey. In this case, the fundamental outsourcing decision has already been taken and customers have to evaluate whether to switch providers. (b) A service is not currently outsourced, but provided in-house.

In this case, a strategic component is added, namely the decision to outsource further logistics tasks or not. While the switching decision (a) should be well captured by the factors contained in the model, strategic outsourcing considerations (b) are not reflected. In fact, this is supported by the findings of Weber and Wallenburg (2004, p. 43), who find that a similar model provides considerably better explanation of additional purchase intentions, when firms intend to increase the amount of outsourced logistics.

Examining the proposed hypotheses reveals that proactive improvement (H_{D13b}) has a particularly strong effect on additional purchase intentions. Thus, firms are willing to entrust their LSPs with new tasks, if LSPs demonstrate continuous improvement efforts. Again, this provides strong evidence of the beneficial signaling effect created by proactive improvement efforts, expressed in a customer's belief that its LSP will excel also in further projects. The second factor to foster additional purchase intentions is commitment (H_{D19b}). As with repurchase intentions, customers are prone to consign additional projects to the currently most important LSP, if they are affectionately committed to the LSP. In addition to the direct effects, both factors display very strong total effects on additional purchase intentions, stressing their overall significance in additional purchasing considerations.

Even though a significant direct effect of relational satisfaction on additional purchases is confirmed by the data, the relationship is negative and does therefore not support H_{D9b}. While this relationship is surprising, the direct effect is equalized by the total effect, which is negligible at 0.02. In order to substantiate this finding, the process of evaluating relational satisfaction should be considered. Relational satisfaction is supposedly closely tied to positive experiences during working relationships between the individuals involved on both sides of a relationship. Since additional outsourcing projects are likely to be handled by a different set of individuals with either party, past relational satisfaction is rendered irrelevant, which provides reasoning for the finding that relational satisfaction is important for relationship maintenance, but not for expansion considerations.

Direct influences of service quality (H_{D1b}), personal trust (H_{D20b}), and alternatives (H_{D25b}) on additional purchase intentions cannot be observed. However, considerable mediated effects that correspond to the surmised coherences exist and thus endorse these factors' relevance to the issue of additional purchase intentions. Consequently, even though the hypothesized direct relationships cannot be confirmed, the underlying effects of service quality, personal trust, and alternatives on additional purchase intentions can still be observed.

Finally, price satisfaction is neither found to directly affect additional purchase intentions (H_{D7b}), nor does a substantial total effect surface. Following the reasoning provided in the preceding section on price satisfaction's influence on repurchase intentions, it can be assumed that prices merely have to meet certain expectations, but do not motivate purchasing decisions that entail existing service providers. In fact, this corresponds to the results obtained by Stank et al. (2003, p. 44), who find that an LSP's cost performance is only a weak predictor of satisfaction. To them, cost performance is merely an order qualifying criterion, while relational performance for instance is an order winning criterion. They caution, however, that "service providers must [...] meet customer expectations at a rate commensurate with delivered value." (p. 44) Information on the hypothesis-tests regarding direct effects of loyalty determinants on additional purchase intentions is reported in Table 7-5.

Table 7-5. Hypothesis-tests for effects on additional purchases (Germany)

Effects of \ Effects on	Additional Purchases	
Service Quality	H_{D1b}	⦸
Price Satisfaction	H_{D7b}	⦸
Relational Satisfaction	H_{D9b}	⦸
Proactive Improvement	H_{D13b}	✓
Fairness	H_{D16b}	×
Commitment	H_{D19b}	✓
Personal Trust	H_{D20b}	⦸
Organizational Trust	H_{D21b}	×
Alternatives	H_{D25b}	⦸

✓ Hypothesis cannot be rejected
⦸ Hypothesis is rejected
× Hypothesis cannot be tested

In contrast to the effects on repurchase intentions, which are considerably different from Wallenburg's (2004) results, a higher degree of consistency can be acknowledged for effects on additional purchase intentions. The only differences to Wallenburg's study pertain to the direct and total effects of commitment and to the direct effect of relational satisfaction. While the effects of commitment are weaker and the direct effect of relational satisfaction is considerably stronger in Wallenburg's model, all other effects are highly consistent.

7.1.1.4 Effects on referrals

Considering the incomplete conceptualization of referrals found in the literature, and the fact that the model consists only of factors that were presumed to directly affect purchase intentions, the model's explanatory power for referral behavior of 54.3% is good and confirms the results obtained by Wallenburg (2004, p. 237), where the R^2 was found to be 56.0%.

Table 7-6. Hypothesis-tests for effects on referrals (Germany)

Effects of \ Effects on	Referrals	
Service Quality	H_{D1c}	✓
Price Satisfaction	H_{D7c}	✓
Relational Satisfaction	H_{D9c}	⊘
Proactive Improvement	H_{D13c}	✓
Fairness	H_{D16c}	×
Commitment	H_{D19c}	✓
Personal Trust	H_{D20c}	⊘
Organizational Trust	H_{D21c}	×
Alternatives	H_{D25c}	⊘

✓ Hypothesis cannot be rejected
⊘ Hypothesis is rejected
× Hypothesis cannot be tested

Exploring the linkages of the seven potential determinants on referrals unveils that direct effects of service quality (H_{D1c}), price satisfaction (H_{D7c}), proactive improvement (H_{D13c}), and commitment (H_{D19c}) support the proposed hypotheses, as reported in Table 7-6.

As presumed, the perception that an LSP provides good service performance leads to its recommendation. While the direct effect is not exceptionally high, service quality's total effect is very strong and underlines the importance of performance for expressions of loyalty. This finding is well matched by the strong effect that proactive improvement has on referrals. Following the argumentation sufficiently elaborated in the preceding sections, proactive improvement does not only nurture purchase intentions, but also induces firms to recommend their LSPs. The specific importance of this aspect, which is obviously a characteristic of superior LSPs, is further underlined by the strong mediated effect on referrals, which exceeds all other total effects. Similarly, commitment has a notably strong effect on referrals, implying that the disposition to recommend an LSP is increased through affectionate ties between customer and provider.

Consistent with the aforementioned results that limited the relevance of price satisfaction for purchasing intentions, prices likewise only have a small effect on referral behavior (H_{D7c}). In fact, the limited relevance of price satisfaction for customer loyalty is in line with findings from McGinnis, Kochunny, and Ackerman (1995), who maintain that price is subordinate to performance in logistics outsourcing considerations, because "price is unlikely to be a source of competitive advantage in the absence of performance competitiveness." (p. 100)

Regarding relational satisfaction, similar results as those obtained for additional purchase intentions are observed. While the direct effect of relational satisfaction on referrals is significant and negative, not supporting H_{D9c}, its total effect is negative, but negligibly small. This can conceivably also be ascribed to the strong personal ties that effectuate relational satisfaction, which will not necessarily be carried over to the party the recommendation is made to.

Finally, alternatives are not found to have a direct effect on referrals (H_{D25c}), but total effects allude to some importance of the existence of market alternatives for referral behavior. Even though the mediated effect of alternatives on referrals is not particularly high at -0.15, it may be concluded that customers evaluate their LSPs more critically when comparisons to alternatives are feasible, and that the motivation to prefer a specific LSP over others is influenced accordingly.

In contrast to the results presented above, Wallenburg (2004, pp. 243-245) only found significant direct effects on referrals for proactive improvement, commitment, and alternatives. Hence, considerable differences between the two models exist. While service quality, price satisfaction, and relational satisfaction have significant direct effects here, they do not directly influence referrals in Wallenburg's study. Also, commitment in Wallenburg's model has a weaker effect than in the present model, and while the present study does not validate a direct effect of alternatives on referrals, Wallenburg indeed found support for that linkage. Regarding direct effects, the only finding consistent between the two studies is the effect of proactive improvement, which is fairly strong in both cases. Examining total effects, though, reveals more consistencies. In addition to proactive improvement, the total effects of service quality and price satisfaction are largely similar. Regarding relational satisfaction and alternatives, the total effects are significantly stronger in Wallenburg's model, while commitment has a stronger total effect in the present study.

7.1.2 Moderation of the German model

As a result of the deliberations in chapter 4.4, a set of hypotheses was derived regarding the moderating effects of opportunism, relationship age, centralization of logistics decisions, and outsourcing focus on the direct linkages between customer loyalty and its determinants. In order to test these hypotheses, comparative analyses were conducted, results of which will be discussed in the following sections.

7.1.2.1 Results of the moderation of opportunism

For opportunism, hypotheses were proposed regarding the direct effects on customer loyalty of service quality, price satisfaction, relational satisfaction, alternatives, and proactive improvement. In order to test these hypotheses ($H_{M1a\text{-}c}$, $H_{M2a\text{-}c}$, $H_{M3a\text{-}c}$, $H_{M4a\text{-}c}$, $H_{M5a\text{-}c}$), one case from the complete data set was first deleted due to incomplete answers to the indicators measuring opportunism, resulting in a total of 544 cases available for further analyses. Then, the data set was split by the median of the aggregated score on opportunism, which corresponds to an average over the four items included in the final measurement model of opportunism, weighted by rotated factor loadings. As a result of a median of 8.04, all scores smaller than 8.04 were termed 'low opportunism' (272 cases) and all values equal or larger than 8.04 were termed 'high opportunism' (272 cases).

Then, all 15 paths for which hypotheses have been proposed were restricted to equality between the two sub-samples and the estimation of this restricted model was compared to the unconstrained solution. As reported in Table 7-7, the increase of χ^2 induced by the restrictions is significant at the 10% significance level, wherefore further analyses were conducted to pinpoint the location(s) of the difference(s).

Table 7-7. Omnibus-test for opportunism (Germany)

Comparison with unconstrained model	Δdf	$\Delta\chi^2$	p
15 paths to customer loyalty constrained	15	24.042	0.064

Following the iterative approach described in chapter 5.6, all paths for which differences had been hypothesized, were then analyzed one-by-one. As a result, two differences were identified: one pertaining to the path between price satisfaction and referrals, which leads to an increase in χ^2 significant at the 5% level. The second significantly moderated path (10%-level) is that between service quality and additional purchases. Details of the analyses are stated in Table 7-8.

Table 7-8. Details for opportunism (Germany)

Customer loyalty determinant	Repurchases			Additional Purchases			Referrals		
	Δdf	$\Delta\chi^2$	p	Δdf	$\Delta\chi^2$	p	Δdf	$\Delta\chi^2$	p
Relational satisfaction	1	0.050	n.s.	1	0.197	n.s.	1	1.523	n.s.
Price satisfaction	1	2.608	n.s.	1	1.18	n.s.	1	5.310	**
Service quality	1	1.378	n.s.	1	2.96	*	1	2.284	n.s.
Proactive improvement	1	2.059	n.s.	1	0.006	n.s.	1	0.655	n.s.
Alternatives	1	1.737	n.s.	1	1.243	n.s.	1	0.571	n.s.
n.s.;*/**/***/****	not significant; significant at the 10%/ 5%/ 1%/ 0.1% level								

Thus, hypotheses H_{M1a-b}, $H_{M2a, c}$, H_{M3a-c}, H_{M4a-c}, and H_{M5a-c} have to be rejected. Examination of the coefficient[1] of the moderated path between price satisfaction and referrals reveals that the relationship is stronger, when opportunism is high, than when it is low, which supports the respective hypothesis H_{M1c}.

According to H_{M2b}, the relationship between service quality and additional purchases was surmised to be stronger, when opportunism is high. This, however, is not supported by the data, because both for low and for high opportunism, the path coefficient is not significant, such that H_{M2b} has to be rejected.

Table 7-9. Paths moderated by opportunism (Germany)

Effect of	Effect on	Opportunism	
		Low	High
Price satisfaction	Referrals	n.s.	0.272****
Service quality	Additional purchases	n.s.	n.s.
n.s.;*/**/***/****	not significant; significant at the 10%/ 5%/ 1%/ 0.1% level		

It is interesting to find that the effect of price satisfaction on referrals is not significant at low levels of opportunism, indicating that prices do not influence a customer's propensity to recommend its LSP. On the other hand, the linkage is highly significant and fairly strong if the LSP is perceived to be opportunistic. Even though the general importance of prices for the formation of customer loyalty is very low – as sufficiently elaborated before – prices do have a decisive impact on referral behavior when the LSP is suspected to be behaving opportunistically. This probably not only reflects the emphasis on current and tangible assessments of satisfaction, but also stresses the role price can play in motivating customers to ascribe less value to other issues, if the prices received are very good. In addition, a strong reliance on prices can be interpreted as a manifestation of a short-term orientation, which is likely if a customer feels insecure as a result of its LSP's opportunistic attitude.

7.1.2.2 Results of the moderation of relationship age

In 521 cases, respondents stated information on the duration of their cooperation with their focal LSP, such that these cases are used in the following analyses. According to the median relationship age of 5.0, the data set was

[1] In comparative analyses in this study, unstandardized path coefficients will always be reported, as standardization may over- or understate the respective path coefficient due to moderations on error terms, variances, and covariances, which are not examined in this study.

split into two parts: a low age sub-sample for values of up to 5 (269 cases), and a high age sub-sample for values of over 5 (252 cases).

As hypotheses were stated for all seven determinants that are included in the validated customer loyalty model in Germany, the respective paths were constrained to equality between the two sub-samples, and the estimation was compared to the unconstrained solution, as reported in Table 7-10.

Table 7-10. Omnibus-test for relationship age (Germany)

Comparison with unconstrained model	Δdf	$\Delta\chi^2$	p
21 paths to customer loyalty constrained	21	34.909	0.029

Since the χ^2-difference is significant at the 5% significance level, detailed analyses on the individual paths were conducted. As shown in Table 7-11, six paths are significantly moderated by relationship age, such that the remaining hypotheses $H_{M6a\text{-}c}$, $H_{M7a\text{-}b}$, $H_{M8a\text{-}b}$, $H_{M9a\text{-}c}$, $H_{M11b\text{-}c}$, and $H_{M14a\text{-}c}$ have to be rejected right away. For the six linkages to customer loyalty, for which significant differences were found, further analyses have to be conducted to assess the corresponding hypotheses.

Table 7-11. Details for relationship age (Germany)

Customer loyalty determinant	Repurchases			Additional Purchases			Referrals		
	Δdf	$\Delta\chi^2$	p	Δdf	$\Delta\chi^2$	p	Δdf	$\Delta\chi^2$	p
Commitment	1	3.864	**	1	0.576	n.s.	1	1.796	n.s.
Personal trust	1	0.009	n.s.	1	0.626	n.s.	1	1.722	n.s.
Relational satisfaction	1	1.510	n.s.	1	2.291	n.s.	1	6.716	***
Price satisfaction	1	1.698	n.s.	1	0.108	n.s.	1	0.007	n.s.
Service quality	1	0.418	n.s.	1	0.028	n.s.	1	7.913	***
Proactive improvement	1	6.306	**	1	3.000	*	1	5.100	**
Alternatives	1	0.126	n.s.	1	2.144	n.s.	1	0.026	n.s.
n.s.;*/**/***/****	not significant; significant at the 10%/ 5%/ 1%/ 0.1% level								

The effects of relational satisfaction and service quality on referrals are moderated by relationship age at the 1% significance level. The path coefficients (Table 7-12), however, indicate that the linkages are stronger in 'aged' than in 'young' relationships. Since these effects are in conflict with the corresponding hypotheses, which surmised that the impacts should be weaker in older relationships, H_{M7c} and H_{M8c} have to be rejected. One possible explanation for this finding is that at given levels of relational satisfaction and service quality, a customer will be more prone to positively refer an LSP, because the danger of making a wrong recommendation is the smaller, the more relational experience a customer has with an LSP. This reasoning, however, cannot be backed by the theoretical framework of social exchange and related theories applied in this research, wherefore it could not be included in hypothesis development. Similarly, the effect of commitment on repurchase intentions was hypothesized to be stronger in older relationships, which is not confirmed by the data indicating a weaker effect for 'aged' relationship. Consequently, H_{M11a} is also rejected.

Table 7-12. Paths moderated by relationship age (Germany)

Effect of	Effect on	Relationship age	
		Low	High
Relational satisfaction	Referrals	n.s.	-0.425****
Commitment	Repurchases	0.458****	0.257***
Service quality	Referrals	n.s.	0.466****
Proactive improvement	Repurchases	n.s.	0.206****
	Additional purchases	0.459****	0.263***
	Referrals	0.465****	0.206**
n.s.;*/**/***/****	not significant; significant at the 10%/ 5%/ 1%/ 0.1% level		

The hypotheses regarding the relationship between proactive improvement and customer loyalty, however, are undirected and H_{M13a}, H_{M13b}, and H_{M13c} can therefore not be rejected. The effect of proactive improvement on repurchase intentions is not significant for 'young' relationships, while it is highly significant in 'aged' relationships. In contrast to that, the effects of proactive improvement on additional purchase intentions and referral behavior are highly significant and fairly strong in 'young' relationships, while they are weaker in 'aged' relationships. Therefore, the importance of proactive improvement for repurchase intentions can be concluded to be stronger in 'aged' than in 'young' relationships, while it is lower for effects on additional purchase intentions and referrals. In the considerations leading to H_{M13a-c}, a direction of the effect could not be argued due to the duality of proactive improvement's consequences. While the shorter-term, performance-increasing effect was expected to be weaker in 'aged' relationships, attributable to the decreased importance of perceptions of satisfaction in these relationships, the longer-term signaling effect was proposed to be stronger in 'aged' relationships. The obtained finding supports the ambiguous argumentation – for repurchase considerations, relationship age appears to affect the long-term signaling effect, while it seems to influence the short-term, performance-improving effect in additional purchase and referral considerations.

7.1.2.3 Results of the moderation of centralization

Prior to conducting moderation analyses for the centralization of logistics decision, one case was eliminated from the data set, because information on centralization was not provided. The resulting sample of 544 cases was split into two parts by its median of 6.0, leading to one sub-sample for low centralization at scores below 6 (258 cases), and a second sub-sample for high centralization at scores of at least 6 (286 cases).

Table 7-13. Omnibus-test for centralization (Germany)

Comparison with unconstrained model	Δ**df**	$\Delta\chi^2$	**p**
21 paths to customer loyalty constrained	21	24.928	0.25

Hypotheses were stated for the loyalty-effects of all determinants contained in the German customer loyalty model, wherefore these paths (21) were constrained to equality and compared to the unconstrained solution. As reported in Table 7-13, imposing these equality constraints does not significantly impair χ^2. A moderation of the respective linkages can therefore not be attested and all hypotheses ($H_{M15a\text{-}c}$, $H_{M16\text{-}c}$, $H_{M17a\text{-}c}$, $H_{M18a\text{-}c}$, $H_{M19a\text{-}c}$, $H_{M22a\text{-}c}$, $H_{M23a\text{-}c}$) are rejected.

As a result of a differing intensity of interaction with the LSP and of differing levels of market knowledge, it was surmised that a centralization of logistics decisions would lead to a focus on easily assessable factors such as service quality and price satisfaction, with relational factors such as commitment and trust loosing importance. These expectations, however, are not supported by the data and the formation of customer loyalty shows to be independent from the centralization of logistics decisions. This leads to the conclusion that logistics decision making follows the same mechanism, whether taken centrally or decentrally and that the employed loyalty determinants are equally relevant in both settings.

7.1.2.4 Results of the moderation of outsourcing focus

The outsourcing focus was determined using the procedure outlined in chapter 6.3.3 and, because of incomplete answers or ambiguous outsourcing foci, 20 cases had to be excluded from the analyses. The remaining 525 cases were split up between externally focused arrangements (480 cases) and internally focused arrangements (45 cases).

Since hypotheses were posited regarding the loyalty-effects of service quality, price satisfaction, relational satisfaction, personal trust, and alternatives, the corresponding paths (15) were constrained to equality and compared to the unconstrained solution in an omnibus-test. As reported in Table 7-14, model adaptation is significantly inferior for the constrained solution, indicating that the examined paths are moderated by the outsourcing focus.

Table 7-14. Omnibus-test for outsourcing focus (Germany)

Comparison with unconstrained model	Δdf	$\Delta\chi^2$	p
15 paths to customer loyalty constrained	15	29.819	0.013

Consequently, further analyses were conducted to identify moderated paths. As indicated in Table 7-15, six out of the 15 examined paths are actually moderated – hypotheses regarding the nine unmoderated paths (H_{M24c}, H_{M26a-c}, H_{M27a-c}, H_{M28a-b}) have to be rejected right away.

Table 7-15. Details for outsourcing focus (Germany)

Customer loyalty determinant	Repurchases			Additional Purchases			Referrals		
	Δdf	$\Delta\chi^2$	p	Δdf	$\Delta\chi^2$	p	Δdf	$\Delta\chi^2$	p
Personal trust	1	2.899	*	1	4.778	**	1	1.952	0.162
Relational satisfaction	1	6.279	**	1	4.659	**	1	8.655	***
Price satisfaction	1	1.819	n.s.	1	0.751	n.s.	1	2.220	n.s.
Service quality	1	0.955	n.s.	1	0.184	n.s.	1	5.965	**
Alternatives	1	0.003	n.s.	1	0.765	n.s.	1	0.162	n.s.
n.s.;*/**/***/****	not significant; significant at the 10%/ 5%/ 1%/ 0.1% level								

It shows that the linkages between service quality and referrals, relational satisfaction and repurchases, additional purchases, and referrals, and personal trust and repurchases and additional purchases are significantly moderated by the outsourcing focus. However, closer scrutiny reveals that not all linkages reflect the a priori hypothesized moderations (Table 7-16). Specifically, the linkages between relational satisfaction and additional purchase intentions, and between personal trust and repurchase intentions differ from the expected effects, wherefore the two respective hypotheses (H_{M24a}, H_{M29b}) are rejected. The remaining four moderations provide support for the expected effects and hypotheses H_{M24b}, H_{M28c}, H_{M29a}, H_{M29c} cannot be rejected.

It was expected that customers would ascribe less importance to satisfaction in internally focused projects, because these can be expected to be well established, grounded on a multitude of satisfactory experiences in the past. In such relationships, temporary dissatisfaction can be put into perspective and does not lead to severe consequences. This surmise is endorsed by the moderations of the linkages between service quality and referrals, and between relational satisfaction and repurchase intentions and referrals – these paths are weaker in internally than in externally focused relationships between LSPs and their customers. While the same argumentation was proposed for the effects of price satisfaction on loyalty, no moderations were identified, most probably attributable to the already minute loyalty-effects of price.

Table 7-16. Paths moderated by outsourcing focus (Germany)

Effect of	Effect on	Outsourcing focus	
		External	Internal
Service quality	Referrals	0.254**	n.s.
Relational satisfaction	Repurchases	0.151**	n.s.
	Additional purchases	-0.332***	-3.466***
	Referrals	-0.246***	n.s.
Personal trust	Repurchases	0.210***	n.s.
	Additional purchases	n.s.	1.378**
n.s.;*/**/***/****	not significant; significant at the 10%/ 5%/ 1%/ 0.1% level		

Even though no conclusions can be drawn on the influences of personal trust on repurchase intentions and referrals, trust gains importance with regard to additional purchase intentions. While the effect is not significant in externally focused outsourcing relationships, it is significant and extremely strong in internally focused relationships, stressing the beneficial effects of trust on a customer's intentions to expand business with its focal LSP beyond the currently rendered services.

7.1.2.5 Summary of moderation analyses for Germany

Overall, only eight out of 72 hypotheses on moderating effects found support in the German data. Hence, the customer loyalty model appears to be very stable concerning different relational aspects. An overview of the moderation hypotheses is given in Table 7-17.

One path is moderated by opportunism, three by relationship age, and four by the outsourcing focus, while no differences in the constitution of customer loyalty are observed between centralized and decentralized decision structures regarding logistics outsourcing.

However, conclusive power of the single moderation by opportunism is limited. While it was shown that price satisfaction has a greater influence on referrals in older than in young relationships, neither the effects of price satisfaction on the other two dimensions of customer loyalty, nor the effects of other determinants on referrals are moderated by opportunism. Hence, the identified moderation does not appear to be systematic and it is therefore difficult to infer generalizeable conclusions on the moderating effects of opportunism on the constitution of customer loyalty.

Table 7-17. Overview of moderation hypothesis-tests (Germany)

Moderator	Customer loyalty determinant	Repurchases		Additional Purchases		Referrals	
Opportunism	Service quality	H_{M2a}	⊘	H_{M2b}	⊘	H_{M2c}	⊘
	Price satisfaction	H_{M1a}	⊘	H_{M1b}	⊘	H_{M1c}	✓
	Relational satisfaction	H_{M3a}	⊘	H_{M3b}	⊘	H_{M3c}	⊘
	Alternatives	H_{M5a}	⊘	H_{M5b}	⊘	H_{M5c}	⊘
	Proactive improvement	H_{M4a}	⊘	H_{M4b}	⊘	H_{M4c}	⊘
Relationship age	Service quality	H_{M7a}	⊘	H_{M7b}	⊘	H_{M7c}	⊘
	Price satisfaction	H_{M6a}	⊘	H_{M6b}	⊘	H_{M6c}	⊘
	Relational satisfaction	H_{M8a}	⊘	H_{M8b}	⊘	H_{M8c}	⊘
	Commitment	H_{M11a}	⊘	H_{M11b}	⊘	H_{M11c}	⊘
	Fairness	H_{M12a}	×	H_{M12b}	×	H_{M12c}	×
	Personal trust	H_{M9a}	⊘	H_{M9b}	⊘	H_{M9c}	⊘
	Organizational Trust	H_{M10a}	×	H_{M10b}	×	H_{M10c}	×
	Alternatives	H_{M14a}	⊘	H_{M14b}	⊘	H_{M14c}	⊘
	Proactive improvement	H_{M13a}	✓	H_{M13b}	✓	H_{M13c}	✓
Centralization	Service quality	H_{M16a}	⊘	H_{M16b}	⊘	H_{M16b}	⊘
	Price satisfaction	H_{M15a}	⊘	H_{M15b}	⊘	H_{M15b}	⊘
	Relational satisfaction	H_{M19a}	⊘	H_{M19b}	⊘	H_{M19b}	⊘
	Commitment	H_{M17a}	⊘	H_{M17b}	⊘	H_{M17b}	⊘
	Fairness	H_{M20a}	×	H_{M20b}	×	H_{M20b}	×
	Personal trust	H_{M18a}	⊘	H_{M18b}	⊘	H_{M18b}	⊘
	Organizational Trust	H_{M21a}	×	H_{M21b}	×	H_{M21b}	×
	Alternatives	H_{M22a}	⊘	H_{M22b}	⊘	H_{M22b}	⊘
	Proactive improvement	H_{M23a}	⊘	H_{M23b}	⊘	H_{M23b}	⊘
Outsourcing focus	Service quality	H_{M28a}	⊘	H_{M28b}	⊘	H_{M28c}	✓
	Price satisfaction	H_{M27a}	⊘	H_{M27b}	⊘	H_{M27c}	⊘
	Relational satisfaction	H_{M29a}	✓	H_{M29b}	⊘	H_{M29c}	✓
	Fairness	H_{M30a}	×	H_{M30b}	×	H_{M30c}	×
	Personal trust	H_{M24a}	⊘	H_{M24b}	✓	H_{M24c}	⊘
	Organizational Trust	H_{M25a}	×	H_{M25b}	×	H_{M25c}	×
	Alternatives	H_{M26a}	⊘	H_{M26b}	⊘	H_{M26c}	⊘

✓ Hypothesis cannot be rejected
⊘ Hypothesis is rejected
× Hypothesis cannot be tested

In contrast to that, all three loyalty-effects of proactive improvement are significantly moderated by relationship age. However, results show that the influence on repurchase intentions increases with relationship age, while it decreases for the effects on additional purchase intentions and referrals. This reflects the ambiguous argumentation provided during hypothesis development, where it was impossible to determine whether proactive improvement's long-term signaling effect or its short-term performance-increasing effect would prevail in the moderation. Hence, moderation findings do not help to identify the exact mechanisms underlying proactive improvement and results remain inconclusive.

On the other hand, there is some indication that the mechanisms underlying customer loyalty differ depending on whether a logistics service is rendered for inbound or outbound streams, or for internal, value creation activities. Here, the loyalty-effects of relational satisfaction appear to be especially affected, because its effects on repurchase intentions and referrals are not significant in internal projects, while they are highly significant for external projects.

Overall, the German customer loyalty model is only subject to very few moderating effects. While some individual paths are moderated by different relationship characteristics, the entirety of moderations is not systematic and it is impossible to draw general conclusions. Rather, the customer loyalty model appears to be very robust against moderating effects, reflecting that the model in itself is generally valid and that the formation of customer loyalty is hardly affected by contingencies.

7.2 Results from the USA

While the last sections presented analyses of the data collected in Germany, the present chapter will examine the US data set. Following the same structure, a customer loyalty model for the USA will be developed first, before interdependencies between the included determinants and their loyalty-effects will be discussed. Then, results of the moderating analyses on the US data will be reported.

7.2.1 Customer loyalty model in the USA

In the USA, discriminance testing in chapter 6.2.7 led to the elimination of fairness and commitment from the customer loyalty model. Additionally, customer loyalty is represented by one comprehensive construct in the USA – the three dimensions of repurchase and additional purchase intentions and referral behavior were not found to be clearly separable. Consequently, the seven factor model with one outcome construct, as depicted in Figure 7-3, is subjected to empirical analyses.

Results of the SEM estimation for this 7-factor model are reported in Table 7-18 and show that that the model does not satisfactorily represent the data collected in the USA, as expressed by GFI and AGFI. For this reason, model refitting has to be conducted with the aim of obtaining a model that provides better adaptation. However, changing the causal structure of the model by modifying the linkages between the constructs may not be performed, as this would mean sacrificing nomological validity. Thus, model refitting can only be conducted by identifying factors that can be eliminated from the model without substantially reducing the model's explanatory power.

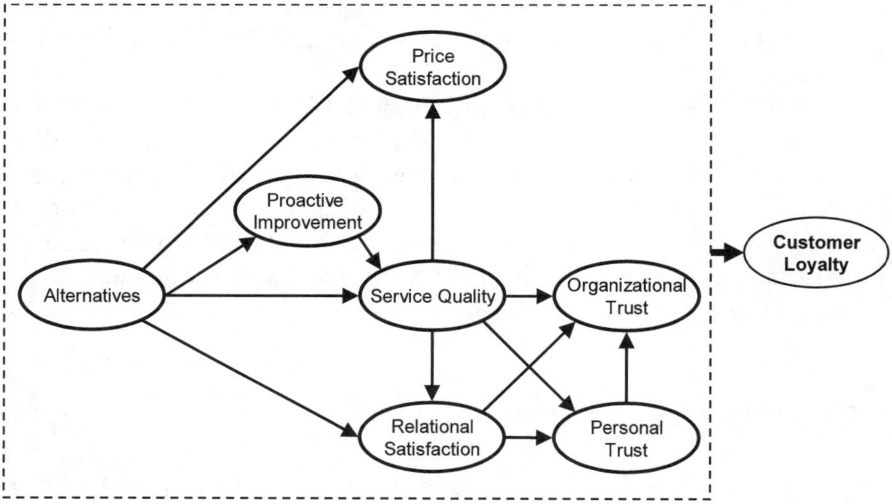

Fig. 7-3. Revised hypothesized model of customer loyalty (USA)

Table 7-18. Adaptation of the 7-factor customer loyalty model (USA)

Adaptation measures						Explanatory power for customer loyalty (R^2)
χ^2/df	TLI	GFI	AGFI	CFI	RMSEA	
1.603	0.967	0.876	0.847	0.971	0.049	71.9%

In a first step, total effects of the seven factors are examined to determine, if all factors provide significant contributions for explaining the loyalty construct (Table 7-19).

This analysis reveals that personal trust, organizational trust, and price satisfaction have total effects on customer loyalty of below 0.100, while all other constructs' effects are considerably stronger. Eliminating these three constructs significantly increases model fit without decreasing the model's explanatory power regarding customer loyalty – the loss of 0.6% percentage points, as reported in Table 7-20, is negligible. Comparing the original 7-factor model with the simplified 4-factor model additionally shows that CAIC and ECVI considerably decreased in the simplified solution, indicating superior model fit.

Table 7-19. Total effects in 7-factor customer loyalty model (USA)

Determinant of Customer Loyalty	Customer Loyalty
	Total Effect
Service Quality	0.754
Relational Satisfaction	0.187
Price Satisfaction	0.013
Proactive Improvement	0.595
Personal Trust	0.089
Organizational Trust	0.015
Alternatives	-0.25
0.65 Standardized total effect	

Table 7-20. Comparison of 7- and 5-factor models (USA)

Model	Adaptation measures								R^2 for customer loyalty
	χ^2/df	TLI	GFI	AGFI	CFI	RMSEA	CAIC	ECVI	
7 factors	1.603	0.967	0.876	0.847	0.971	0.049	959.989	2.548	71.9%
4 factors	1.485	0.983	0.934	0.907	0.987	0.044	403.380	0.894	71.3%

As all fit criteria fulfill the established requirements, the 4-factor model (see Figure 7-4) will be used in the subsequent analyses.

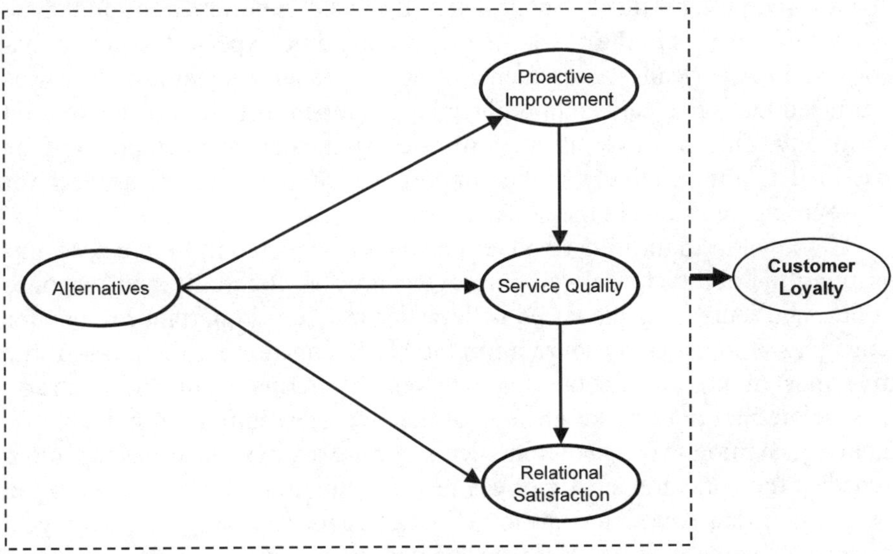

Fig. 7-4. Validated 4-factor customer loyalty model (USA)

This 4-factor model of customer loyalty contains service quality and relational satisfaction as two out of the three dimensions of satisfaction, proactive improvement, and alternatives. Both trust constructs had to be eliminated from the model, because they do not contribute to the explanation of customer loyalty to a significant degree. While this finding is contrary to the findings in the German model, where personal trust is included, it is consistent with the cultural deliberations presented earlier (Chapter 3.4 concluded with an overview of major cultural differences on the grounds of the frameworks by Hofstede 2001, Trompenaars and Hampden-Turner 1997, and Hall and Hall 1990). As stated in Table 3-1, and contrary to Germans, Americans are rather short-term oriented, do not value relationships as highly, and do not strongly require personal bonds for maintaining relationships. Since trust has a long-term effect and evolves out of repeated interaction over time, based on evaluations of credibility and benevolence, the information that trust bears no significance with regard to customer loyalty in the USA is not surprising. This is further supported by the finding that Americans are less prone to extending trust anyway. Additionally, trust's non-significance for customer loyalty in the USA is concurrent with the discovery of Morgan and Hunt (1994), who do not identify a direct effect of trust on retention in their US sample.

The revelation that price satisfaction is not important for the formation of loyalty, however, comes as a surprise. While price satisfaction in included in the German model, even though only having minor implications for loyalty, cultural theory suggests an increased importance of immediate economic benefits in the USA. As prices should be expected to play a major role in Americans' assessments of economic advantageousness, it was surmised that price satisfaction played a prominent role in the constitution of loyalty. This, however, is not confirmed by the data and can probably be ascribed to the relatively higher importance of performance aspects for outsourcing decisions in the USA.

As shown in Figure 7-5 (the German data is depicted in Figure 2-4), improving the ability to deliver, improving process flexibility and response times, and increasing the speed of logistics are very important reasons for engaging in logistics outsourcing in the USA, constituting three out of the five most important outsourcing motives. In Germany, on the contrary, cost-related motives make up four of the five most important outsourcing goals. Assuming that outsourcing decisions are always taken for economic reasons, the comparison of motives in Germany and the USA leads to the conclusion that Americans attribute higher economic potential to the performance increases achievable through logistics outsourcing, while Germans strongly focus on cost aspects. As a result, it is comprehensible that prices play a minor role in the USA compared to Germany, and are therefore not part of the customer loyalty model.

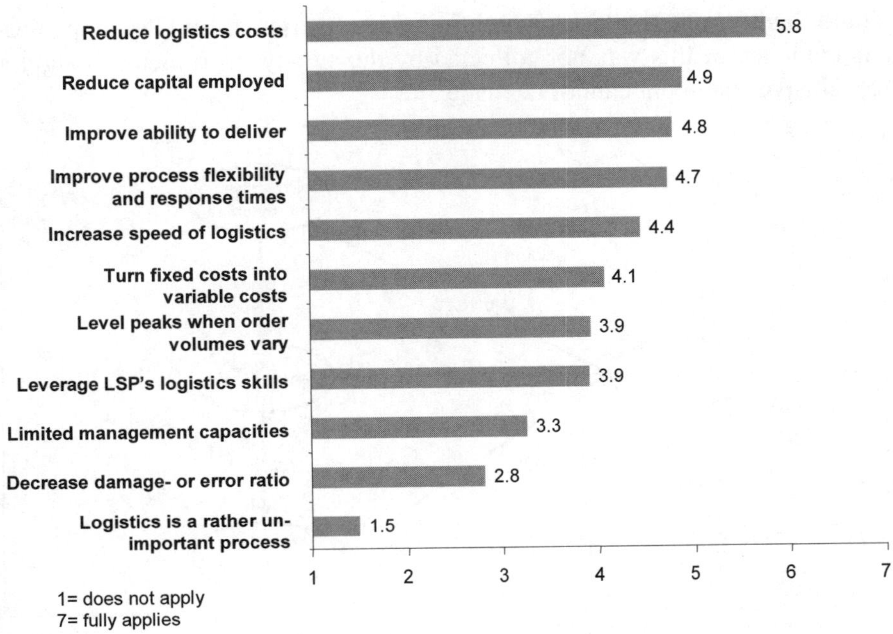

Fig. 7-5. Motivation to outsource in 2005 (USA)

7.2.1.1 Relationships between the determinants

The simplified US model contains a total of five linkages between the four determinants of customer loyalty, as graphically shown in Figure 7-6. Of these, four linkages are significant, supporting hypotheses H_{D5}, H_{D14}, H_{D26}, H_{D27}. The hypothesis stating a direct influence of alternatives on relational satisfaction (H_{D29}) does not find support. Results of the hypotheses-tests are reported in Table 7-21.

As expected, the availability of adequate market alternatives to the currently used LSP has a negative impact on proactive improvement and on service quality. Contrary to the conclusions drawn from social exchange theory, the perception of relational satisfaction is not influenced by alternatives. While Germans' evaluations of relational satisfaction were found to be influenced by the availability of alternatives, Americans evidently appraise relational satisfaction with one LSP without recurring to other LSPs. Taking into consideration that judgments of relationship interaction strongly allude to personal experiences between individuals, the establishment of a comparison level for relational satisfaction would require intimate experiences with other LSPs. A possible explanation for the non-significance of the linkage between alternatives and relational satisfaction could therefore be the lack of such intimate experiences in US companies. As evidence of this was not collected by the survey instrument, though, a conclusive statement cannot be made.

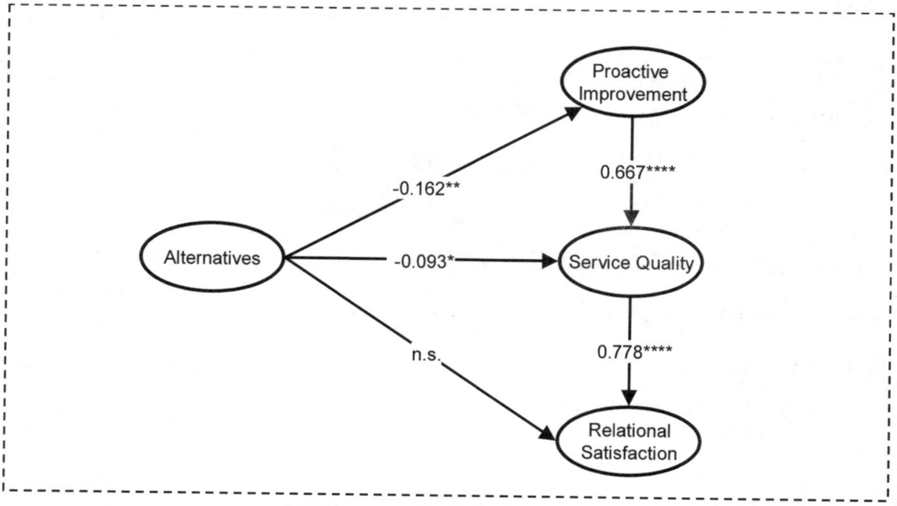

0.655**	Standardized path coefficient with significance level
*/**/***/****	10%/5%/1%/0.1% significance level
n.s.	not significant

Fig. 7-6. Relationships between customer loyalty determinants (USA)

Table 7-21. Hypothesis-tests for linkages between determinants (USA)

Effects of \ Effects on	Service Quality	Price Satisfaction	Relational Satisfaction	Proactive Improvement	Fairness	Commitment	Personal Trust	Organizational Trust
Service Quality		H_{D4} ×	H_{D5} ✓		H_{D6} ×	H_{D2} ×	H_{D3a} ×	H_{D3b} ×
Price Satisfaction					H_{D8} ×			
Relational Satisfaction					H_{D12} ×	H_{D10} ×	H_{D11a} ×	H_{D11b} ×
Proactive Improvement	H_{D14} ✓				H_{D15} ×			
Fairness						H_{D17} ×	H_{D18a} ×	H_{D18b} ×
Commitment								
Personal Trust						H_{D23} ×		H_{D22} ×
Organizational Trust						H_{D24} ×		
Alternatives	H_{D27} ✓	H_{D28} ×	H_{D29} ⊘	H_{D26} ✓				

✓ Hypothesis cannot be rejected
⊘ Hypothesis is rejected
× Hypothesis cannot be tested

Finally, the hypotheses on the effect of proactive improvement on service quality (H_{D14}) and of service quality on relational satisfaction (H_{D5})

find strong support in the US data. Both relationships are particularly strong, indicating that the three constructs are closely tied to one another. Proactive improvement is a decisive driver of service quality, which corresponds to the direct performance-increasing effect hypothesized for proactive improvement. Similarly, satisfaction with performance facilitates interaction, as there are fewer conflicts to be resolved, stated in the strong effect of service quality on relational satisfaction.

7.2.1.2 Effects on customer loyalty

Regarding the effects of alternatives, proactive improvement, service quality, and relational satisfaction on customer loyalty, all hypotheses are supported by the data – direct and total effects are reported in Table 7-22 and an overview of hypothesis-tests is provided in Table 7-23. In addition, the model is well suited for explaining customer loyalty, as 71.3% of the variance of the customer loyalty construct is captured by the model, which means that only 28.7% are attributable to influences not modeled.

Table 7-22. Effects of determinants on customer loyalty (USA)

	Customer Loyalty	
	Direct Effect	Total Effect
Service Quality	0.58****	0.74
Relational Satisfaction	0.20**	0.20
Proactive Improvement	0.11*	0.60
Alternatives	-0.07*	-0.25

0.16**	Standardized path coefficient with significance level
0.65	Standardized total effect
n.s.;*/**/***/****	not significant; significant at the 10%/ 5%/ 1%/ 0.1% level

Examining the direct effects first, a particularly strong effect on loyalty can be detected for service quality, which is substantially stronger than all other direct effects. Including total effects in the interpretation of the data, though, reveals that both service quality and proactive improvement are important drivers of loyalty. Again, this underlines the dual effect of proactive improvement, which, in addition to increasing performance, demonstrates LSPs' willingness to excel for the benefit of their customers.

In addition, the minute direct effect of alternatives is put into perspective by viewing total effects. As suggested by theory, the majority of the effect of alternatives on loyalty is mediated – as pointed out in the preceding section – through service quality and proactive improvement. This indicates that competing LSPs do not have a decisive direct impact on loyalty considerations, but that an influence is mainly exerted by serving as comparison standards for the evaluation of other factors affecting loyalty.

Table 7-23. Hypothesis-tests for effects on customer loyalty (USA)

Effects of \ Effects on	Customer Loyalty	
Service Quality	H_{D1}	✓
Price Satisfaction	H_{D7}	×
Relational Satisfaction	H_{D9}	✓
Proactive Improvement	H_{D13}	✓
Fairness	H_{D16}	×
Commitment	H_{D19}	×
Personal Trust	H_{D20}	×
Organizational Trust	H_{D21}	×
Alternatives	H_{D25}	✓

✓ Hypothesis cannot be rejected
⊘ Hypothesis is rejected
× Hypothesis cannot be tested

Relational satisfaction has a fairly strong direct effect on customer loyalty, which is significant at the 5%-level. Since the effect on customer loyalty is the only effect emanating from relational satisfaction, the total effect is equal to the direct effect. A second interesting finding is the very strong total effect of service quality on customer loyalty, indicating that a substantial part of the loyalty-effect of service quality is mediated through relational satisfaction. This further underlines the importance of service quality for relational satisfaction, as posited by social exchange theory.

7.2.2 Moderation of the US model

The following sections will describe the moderating analyses conducted on the US model. For that, the effects of the four relationship characteristics of opportunism, relationship age, centralization of logistics decisions, and outsourcing focus on the linkages between customer loyalty and alternatives, proactive improvement, service quality, and relational satisfaction will be examined and the respective hypotheses will be tested and discussed.

7.2.2.1 Results of the moderation of opportunism

Before being able to conduct moderating analyses on opportunism, one case in which responses for opportunism were not provided had to be deleted from the data set, resulting in 249 cases usable for the analyses. Subsequently, the data set was split by the median of the aggregated opportunism score into sub-sample 1 for values below 10 (135 cases), indicating low opportunism, and sub-sample 2 for values of at least 10 (114 cases), indicating high opportunism.

Since hypotheses on the moderation by opportunism were proposed for all determinants included in the model, the respective four direct effects on customer loyalty were constrained to equality in both samples and the estimation was compared with the unconstrained solution. As reported in Table 7-24, the resulting impairment of model adaptation is highly significant, wherefore further analyses were conducted.

Table 7-24. Omnibus-test for opportunism (USA)

Comparison with unconstrained model	Δdf	$\Delta\chi^2$	p
4 paths to customer loyalty constrained	4	17.872	0.001

Table 7-25. Details for opportunism (USA)

Customer loyalty determinant	Customer Loyalty		
	Δdf	$\Delta\chi^2$	p
Service quality	1	7.322	***
Relational satisfaction	1	11.369	****
Alternatives	1	1.274	n.s.
Proactive improvement	1	5.655	**
n.s.;*/**/***/****	not significant; significant at the 10%/ 5%/ 1%/ 0.1% level		

The detailed analyses (Table 7-25) reveal that three of the four effects on customer loyalty are significantly moderated by opportunism. Merely the loyalty-effect of alternatives is stable and H_{M5}, which forecasted a stronger effect when opportunism is perceived, is consequently rejected. Whether or not the focal LSP exhibits opportunistic behavior does not influence the way in which the availability of adequate market alternatives exerts a direct influence on customer loyalty. It should be noted, though, that alternatives' direct effect on customer loyalty is very small in the base model, because the major share of the factor's influence is effected through mediation via satisfaction and proactive improvement, where alternatives serve as a comparison standard. Statements on whether the impact of alternatives on other loyalty determinants is moderated by opportunism cannot be made, however, as interdependencies between determinants are not subjected to moderation analyses within this research.

For the remaining, significantly moderated paths, examination of the parameter values shows that two of the three moderations are as hypothesized (Table 7-26). It was surmised that assessments of satisfaction would be more important for customer loyalty under high than under low opportunism, which finds support in the effect of relational satisfaction, but not in the effect of service quality. Therefore, H_{M2} is rejected.

Table 7-26. Paths moderated by opportunism (USA)

Effect of	Effect on	Opportunism	
		Low	High
Service quality	Customer Loyalty	1.054****	0.488****
Relational satisfaction	Customer Loyalty	n.s.	0.544****
Proactive improvement	Customer Loyalty	n.s.	0.171**
n.s.;*/**/***/****	not significant; significant at the 10%/ 5%/ 1%/ 0.1% level		

The effect of relational satisfaction on customer loyalty is actually not significant in the low opportunism sample, while it is significant and strong in the high opportunism sample. As posited, relational satisfaction is more important in customer loyalty considerations, when the LSP is perceived to be behaving opportunistically. Successful interaction between an LSP and its customer can thus function as a means to reduce the detrimental effects of opportunism, alleviating the fear that the LSP could take advantage of its customer.

Finally, an undirected hypothesis was proposed for the moderation on the linkage between proactive improvement and customer loyalty. Tending to the duality of the loyalty-effect of proactive improvement, which both increases service quality and signals an LSP's willingness to excel for its customer's benefit, the argumentation on the moderating effect of opportunism was ambiguous. While service quality was expected to be of more importance, the longer-term signaling effect was suspected to be weaker at high levels of opportunism. As stated by the data, proactive improvement is in fact more important for loyalty under high than under low opportunism, which at first sight stresses the argumentation that the performance increasing effect prevails in the moderation. However, the hypothesis on the effect of service quality on customer loyalty, which is based on the same argumentation, had to be rejected, because the data actually suggests an inverted effect, in which service quality is less important for loyalty under high opportunism. Consequently, the performance increasing effect of proactive improvement can therefore not be expected to be more important in settings where customers are faced with opportunistic LSPs. Even though the data suggests that the effect of proactive improvement is stronger in these circumstances, the evidence is dubious, which is also reflected in the relatively low significance level for the moderated path under high opportunism (5% level) and in the small path coefficient. Thus, a conclusion on the moderating effect of opportunism on the loyalty-effect of proactive improvement will not be drawn.

7.2.2.2 Results of the moderation of relationship age

As 11 respondents did not state the age of the relationship to their focal LSP, these cases had to be eliminated from the data set, leading to 239 cases usable for the moderation analyses of relationship age. Following standard procedure, median splitting was performed to obtain two sub-samples, the first of which includes 131 cases (relationship age of up to four years), and the second of which comprises the remaining 108 cases (relationship age of over four years).

Since hypotheses had been posited for the loyalty-effects of all determinants in the US model, omnibus-testing in which these four paths were constrained was conducted. As shown in Table 7-27, the resulting increase of χ^2 is just significant with a p-value of 0.099, such that further analyses were conducted.

Table 7-27. Omnibus-test for relationship age (USA)

Comparison with unconstrained model	Δdf	$\Delta\chi^2$	p
4 paths to customer loyalty constrained	4	7.796	0.099

Table 7-28. Details for relationship age (USA)

Customer loyalty determinant	Customer Loyalty		
	Δdf	$\Delta\chi^2$	p
Service quality	1	2.284	n.s.
Relational satisfaction	1	5.291	**
Alternatives	1	0.710	n.s.
Proactive improvement	1	1.530	n.s.
n.s.;*/**/***/****	not significant; significant at the 10%/ 5%/ 1%/ 0.1% level		

The iterative analyses on the four individual paths reveal that only the relationship between relational satisfaction and customer loyalty is significantly moderated by the age of the relationship between customers and their focal LSP (Table 7-28). Consequently, hypotheses on the moderation of the loyalty-effects of service quality (H_{M7}), alternatives (H_{M14}), and proactive improvement (H_{M13}) are rejected.

The effect of relational satisfaction on customer loyalty, as exhibited in Table 7-29, is fairly strong and significant at the 1% level in 'young' relationships, while it is not significant in 'aged' relationships. This supports H_{M8}, which can therefore not be rejected. As expected, satisfaction with the interaction with and relationship between an LSP and its customer loses importance in loyalty considerations in 'old' and established relationships.

Table 7-29. Path moderated by relationship age (USA)

Effect of	Effect on	Relationship age	
		Low	High
Relational satisfaction	Customer Loyalty	0.383***	n.s.
n.s.;*/**/***/****	not significant; significant at the 10%/ 5%/ 1%/ 0.1% level		

As customers do not dispose of sufficient relational experience to be able to put particular negative perceptions into perspective in 'young' relationships, their evaluations of relational satisfaction are highly influenced by potential temporary dissatisfaction. In the absence of other relational governance norms, which also need time to develop, low relationship age leads to an increased impact of relational satisfaction on customer loyalty. When relationship age is high, on the contrary, relational satisfaction loses importance, because transient feelings of dissatisfaction can be put into perspective.

7.2.2.3 Results of the moderation of centralization

For analyzing the moderating effect of the centralization of logistics decisions on the US customer loyalty model, one case had to be eliminated, because the respective question was not answered, wherefore 249 cases are available for the analyses. Median splitting resulted in one sub-sample for low centralization (120 cases with scores smaller six), and a second sub-sample for high centralization (119 cases with scores of at least six).

Table 7-30. Omnibus-test for centralization (USA)

Comparison with unconstrained model	Δdf	$\Delta\chi^2$	p
4 paths to customer loyalty constrained	4	4.89	0.299

Hypotheses regarding the moderating effects of centralization were proposed for all four determinants included in the US customer loyalty model and an omnibus-test was therefore conducted, in which the four paths to customer loyalty were constrained to equality between the two sub-samples. As reported in Table 7-30, this does not lead to a significant impairment of model adaptation. Consequently, the data does not suggest that the centralization of logistics decisions has a moderating effect on the direct effects on customer loyalty. Accordingly, the corresponding hypotheses, H_{M16}, H_{M19}, H_{M22}, and H_{M23}, have to be rejected.

7.2.2.4 Results of the moderation of outsourcing focus

Performing the computation described in chapter 6.3.3, the data set is separated into externally focused cases (183), in which the LSP renders inbound and outbound logistics services, and internally focused cases (46), in which services are rendered within the customer's firm boundaries. 21 cases have to be excluded from further analyses, because either no indication of the focus was provided by respondents, or the focus was not clearly internal or external.

Table 7-31. Omnibus-test for outsourcing focus (USA)

Comparison with unconstrained model	Δdf	$\Delta\chi^2$	p
3 paths to customer loyalty constrained	3	3.526	0.317

An omnibus-test was conducted in which the three paths leading to customer loyalty, for which moderation hypotheses had been formulated, were restricted to equality between the internal and external samples. As shown in Table 7-31, the restriction does not significantly worsen model adaptation, denoting that the loyalty-effects of service quality, relational satisfaction, and alternatives are not moderated by the outsourcing focus. Hence, hypotheses H_{M26}, H_{M28}, and H_{M29} have to be rejected.

Even though the outsourcing focus was found to be a moderator of the customer loyalty model in Germany, no such indication is suggested for the USA. While satisfaction was shown to be a subordinate antecedent to customer loyalty in German internally focused projects, US-Americans do not distinguish between internally and externally focused outsourcing projects. Instead, the three determinants of customer loyalty for which a moderation was expected in the USA, i.e. alternatives, service quality, and relational satisfaction, exhibit the same effects on loyalty, no matter where the logistics services are rendered.

7.2.2.5 Summary of moderation analyses for the USA

Table 7-32. Overview of moderation hypothesis-tests (USA)

Moderator	Customer loyalty determinant	Customer loyalty	
Opportunism	Service quality	H_{M2}	⊘
	Price satisfaction	H_{M1}	×
	Relational satisfaction	H_{M3}	✓
	Alternatives	H_{M5}	⊘
	Proactive improvement	H_{M4}	✓
Relationship age	Service quality	H_{M7}	⊘
	Price satisfaction	H_{M6}	×
	Relational satisfaction	H_{M8}	✓
	Commitment	H_{M11}	×
	Fairness	H_{M12}	×
	Personal trust	H_{M9}	×
	Organizational Trust	H_{M10}	×
	Alternatives	H_{M14}	⊘
	Proactive improvement	H_{M13}	⊘
Centralization	Service quality	H_{M16}	⊘
	Price satisfaction	H_{M15}	×
	Relational satisfaction	H_{M19}	⊘
	Commitment	H_{M17}	×
	Fairness	H_{M20}	×
	Personal trust	H_{M18}	×
	Organizational Trust	H_{M21}	×
	Alternatives	H_{M22}	⊘
	Proactive improvement	H_{M23}	⊘
Outsourcing focus	Service quality	H_{M28}	⊘
	Price satisfaction	H_{M27}	×
	Relational satisfaction	H_{M29}	⊘
	Fairness	H_{M30}	×
	Personal trust	H_{M24}	×
	Organizational Trust	H_{M25}	×
	Alternatives	H_{M26}	⊘

✓ Hypothesis cannot be rejected
⊘ Hypothesis is rejected
× Hypothesis cannot be tested

As demonstrated in the previous sections, only three out of the 15 moderation hypotheses find support for the US data. While the centralization of logistics decisions and the outsourcing focus do not have any moderating effects on the US model, some moderation is observed for opportunism and relationship age. Table 7-32 provides an overview of all hypotheses tested on the US sample.

While the effects of relational satisfaction and proactive improvement on customer loyalty are both stronger when a customer deals with an opportunistic LSP, the latter effect is dubious, as pointed out earlier. Even though service quality does not have a greater impact on customer loyalty under opportunism, proactive improvement does. This, however, contradicts the argumentation provided in chapter 4.4.2.1, according to which the future-oriented signaling effect caused by proactive improvement should be limited by opportunism. It was surmised that customers would not be confident that opportunistic LSPs would be innovative and exceed their contractual obligations by proactively pursuing improvements for the customers' benefit. Nevertheless, this argumentation does not seem to hold, as customers faced with opportunistically behaving LSPs actually do emphasize proactive improvement in their loyalty considerations. While this may be an interesting revelation, its conclusive power is limited by the relatively low parameter coefficient between proactive improvement and loyalty, paired with a significance level of only 5%.

Overall, the empirical evidence provides very little indication that the model is susceptible to influences outside the modeled determinants. The degree of centralization of logistics decisions and the outsourcing focus do not affect the model at all, and opportunism and relationship age do not appear to have strong effects either. As such, the US data supports the reasoning provided for moderations on the German model. The major implication from testing for moderating influences therefore is that it is impossible to infer general conclusions from the empirical evidence. Rather, both the German and the US data suggest that the model is mainly robust, which also stresses the general applicability of the theories employed to derive the conceptual model.

7.3 Differences between Germany and the USA

The preceding sections developed customer loyalty models for the USA and Germany, which are considerably different. The German model includes seven loyalty determinants (alternatives, proactive improvement, price satisfaction, service quality, relational satisfaction, commitment, and personal trust) that affect three dimensions of customer loyalty (repurchase and additional purchase intentions, referral behavior). Meanwhile, the US model consists of only 4 loyalty determinants (alternatives, proactive improvement, service quality, and relational satisfaction) and one comprehensive loyalty construct. For this reason, quantitative evaluation of the hypotheses presented in chapter 4.5.1 (see the overview in Table 4-3) is not possible. Deriving meaningful results from comparative analyses in structural equation modeling requires a high degree of similarity of the two models under scrutiny, a requirement which is not satisfied by the German and US models.

Even though hypothesis testing for culture-induced differences can therefore not be performed, differences between the two models will be discussed in the following. For that purpose, the subsequent sections will first discuss differences between Germany and the USA with regard to the dimensionality of customer loyalty, before differences with regard to loyalty effects of service quality, price satisfaction, relational satisfaction, personal trust, and alternatives will be presented.

Fairness, commitment, and organizational trust cannot be analyzed, as these constructs were excluded from analyses in Germany and/or the USA as a result of lacking discriminant validity.

7.3.1 Dimensionality of customer loyalty

The first substantial difference between Germany and the USA is the differing dimensionality of customer loyalty. Following the conceptualization of customer loyalty presented in chapter 2.1.2.3, three distinct customer loyalty dimensions were derived and corresponding constructs were developed. While this three dimensional conceptualization originally stems from German marketing research, going back to an article by Meyer and Oevermann (1995), and later adopted by such authors as Homburg and Fassnacht (2001), Braunstein (2001), and Wallenburg (2004), a similar conceptualization of customer loyalty is not found in studies with non-German data. Rather, different combinations of the three dimensions are employed in international research, with articles focusing either on repurchase and additional purchase intentions (e.g. Homburg, Giering, and Menon 2003), on repurchase intentions and referrals (e.g. Bloemer, de Ruyter, and Wetzels 1999, Sivadas and Baker-Prewitt 2000, Knemeyer and Murphy 2005), or on any dimension of loyalty individually. Although this suggests that a conceptual distinction between different loyalty dimensions is generally acknowledged, evidence of the applicability of a multi-dimensional customer loyalty conceptualization, incorporating both purchasing intentions and referral behavior, to US data is not provided. A first explanation for the uni-dimensionality of customer loyalty in the USA is therefore a potential difference in the notion of customer loyalty.

The measurement of customer loyalty performed in this study could be another explanation for the differing dimensionality of customer loyalty. As reported in chapter 6.1, measurement was exclusively based on scales developed by Giering (2000) and Wallenburg (2004), who operationalized the three constructs with a German bias. Such a cultural bias is inevitable, as researchers ineluctably develop scales against the background of their cultural descent. While this is useful in research, where scales are only applied to one culture alone, scales may not be transferable to other cultural settings. Thus, the measurement models used in the present study may just not capture all relevant aspects of the three dimensions of customer loyalty that would make them unique and separable also in the USA.

The third possible explanation for the differing dimensionality of customer loyalty is methodological. Assessment of the adaptation of latent variables in this study was based on the two-step procedure proposed by Homburg and Giering (1996), in which both exploratory and confirmatory factor analysis are applied. Similarly, the validity of the three customer loyalty dimensions was evaluated using EFA and CFA. The Fornell/Larcker criterion, which assesses discriminant validity by juxtaposing constructs' squared correlations to average variances extracted, actually showed discriminance. EFA, however, which can be used to assess convergent validity, revealed that all indicators are highly correlated, loading on only one construct. On the basis of this finding, the very stringent decision was taken to treat loyalty in the USA as one construct, comprising different aspects of purchasing intentions and referral behavior. While this approach is very rigorous, information on EFA is not commonly reported in relevant US journals and it is therefore not guaranteed that EFA is incorporated in model assessment in the USA. Abstaining from the evaluation of convergent validity, the hypothesized three dimensional customer loyalty conceptualization would in fact find support also in the USA.

7.3.2 Cultural comparison with regard to service quality

Hypothesis $H_{C1a\text{-}c}$ states that the effect of service quality on customer loyalty should be stronger in the USA than in Germany. This is based both on considerations based on the three addressed cultural frameworks, as well as on previous research that dealt with service quality in international settings. As reported in Table 7-33, service quality in fact has a particularly strong effect of 0.58 in the USA. In Germany, direct effects on repurchases and referrals are considerably weaker and no significant impact on additional purchases was found.

Table 7-33. Comparison Germany-USA for service quality

Hypothesis	Loyalty Determinant	Germany			USA
		Effects on			Effect on
		Repurchases	Additional purchases	Referrals	Customer loyalty
H_{C1a-c}	Service quality	0.16**	n.s.	0.19**	0.58****
	0.16**	Standardized path coefficient with significance level			
	n.s.;*/**/***/****	not significant; significant at the 10%/ 5%/ 1%/ 0.1% level			

Regarding total effects, the differences in strength are not as pronounced In Germany, service quality has a total effect of 0.65/0.18/0.41 on repurchases/additional purchases/referrals, while the total effect on customer loyalty in the US is 0.74. However, the total effect in the USA is still considerably stronger than the German total effects, wherefore some support is provided for the hypothesized stressed importance of service quality in the USA.

7.3.3 Cultural comparison with regard to price satisfaction

Table 7-34. Comparison Germany-USA for price satisfaction

Hypothesis	Loyalty Determinant	Germany			USA
		Effects on			Determinant not included in the model
		Repurchases	Additional purchases	Referrals	
H_{C2a-c}	Price satisfaction	0.05*	n.s.	0.09***	
	0.16**	Standardized path coefficient with significance level			
	n.s.;*/**/***/****	not significant; significant at the 10%/ 5%/ 1%/ 0.1% level			

All three cultural frameworks agree that short-term economic benefits are important in the USA and that Americans are more willing to switch providers if economic inequities are observed in the short run. This lead to hypothesis H_{C2a-c} which states that price satisfaction should have a stronger effect on loyalty in the USA than in Germany. Examining the two models, however, does not support this surmise at all. While price satisfaction is only of minor importance in loyalty considerations in Germany, the construct was eliminated from the US model on the grounds of its negligibly small impact on loyalty. As pointed out before, the reason for this may be found in differing assessments of economic benefits in Germany and the USA.

While hypothesis formulation assumed that prices would decisively influence the evaluation of economic beneficialness, Americans appear to place great value on performance aspects. This was already sufficiently elaborated before, when the comparison of Germans' and Americans' outsourcing motivations revealed that performance aspects play a more pronounces role in outsourcing decisions in the USA. Overall, prices only appear to be order qualifying criteria in both countries.

7.3.4 Cultural comparison with regard to relational satisfaction

Table 7-35. Comparison Germany-USA for relational satisfaction

Hypothesis	Loyalty Determinant	Germany			USA
		Effects on			Effect on
		Repurchases	**Additional purchases**	**Referrals**	**Customer loyalty**
H_{C3a-c}	Relational satisfaction	0.15**	-0.22**	-0.17**	0.20**
	0.16**	Standardized path coefficient with significance level			
	n.s.;*/**/***/****	not significant; significant at the 10%/ 5%/ 1%/ 0.1% level			

Hypothesis H_{C3a-c} proposes that the relational satisfaction-loyalty linkage would be stronger in Germany than in the USA. As Germans dread conflict and avoid ambiguity, seeking mutually beneficial cooperation, they value good interpersonal relationships. The American culture, on the other hand, is rather impersonal, and economic reasoning overrules the importance of relationships.

The hypothesized effect, however, does not surface in the data. While ambiguous results were obtained in Germany for relational satisfaction's effects on additional purchases and referrals – as discussed earlier – the strength of the effect on repurchase intentions is comparable to the effect on the customer loyalty construct in the USA. Without algebraic hypothesis testing, a difference between Germany and the USA cannot be derived from the empirical evidence.

While this finding is clearly contradicting the intercultural research provided in chapter 3.4, it should be noted that after all, German and American culture were presented to display more similarities than differences and that the proposed hypothesis may have overstated the actual difference. Furthermore, the relational satisfaction construct referred to an affective and cognitive evaluation of all experiences made by a customer with the interaction with and relationship to an LSP. Consequently, German and American respondents may have incorporated different standards in this evaluation, leading to an aetiologically different perception of relational satisfaction that disguises potential differences in its constituting forces.

7.3.5 Cultural comparison with regard to personal trust

Table 7-36. Comparison Germany-USA for personal trust

Hypothesis	Loyalty Determinant	Germany			USA
		Effects on			Determinant not included in the model
		Repurchases	Additional purchases	Referrals	
$H_{C6a\text{-}c}$	Personal trust	0.17***	n.s.	n.s.	
	0.16**	Standardized path coefficient with significance level			
	n.s.;*/**/***/****	not significant; significant at the 10%/ 5%/ 1%/ 0.1% level			

Hypothesis $H_{C6a\text{-}c}$ suggests that the linkage between personal trust and loyalty is more pronounced in Germany than it is in the USA. Characteristic of collectivist societies, cultural theory maintains that long-term business relationships in Germany require personal bonding and trust for relationship maintenance. As a result, private relations often evolve and individuals of both parties admit the respective other side to their in-groups. This is not the case in the USA, where in- and out-groups are not distinguished, personal bonding in the business world is rare, and trust is not a prerequisite for relationship maintenance.

The German customer loyalty model states a positive relationship between personal trust and repurchase intentions, but not between personal trust and the other two loyalty dimensions. This is consistent with cultural theory, because repurchase intentions are concerned with the continuation of an existing relationship with a fixed set of individuals. Additional purchases and referrals, on the other hand, do not necessarily involve the same individuals that are engaged in the current relationship and personal trust, which is tied to specific individuals, is not as important (see also sections 7.1.1.3 and 7.1.1.4). In the USA, personal trust is not even included in the customer loyalty model, resulting from the weak influence on customer loyalty revealed in the initial analyses. Thus, H_{C6a} and the expectation that trust does not influence Americans in their loyalty considerations are supported by the data.

7.3.6 Cultural comparison with regard to alternatives

Table 7-37. Comparison Germany-USA for alternatives

Hypothesis	Loyalty Determinant	Germany			USA
		Effects on			Effect on
		Repurchases	Additional purchases	Referrals	Customer loyalty
$H_{C8a\text{-}c}$	Alternatives	-0.07****	n.s.	n.s.	-0.07*
	0.16**	Standardized path coefficient with significance level			
	n.s.;*/**/***/****	not significant; significant at the 10%/ 5%/ 1%/ 0.1% level			

The existence of adequate alternatives to the currently used LSP, i.e. the availability of alternative LSPs offering similar prices and performance, is a prerequisite for switching. In this respect, cultural theory lead to the supposition that Germans, who dread change and innovation and are generally more loyal to relationships than Americans, have a lower inclination towards switching. This argument is supported by empirical work of Liu, Furrer, and Sudharshan (2001), who observe that members of individualistic societies, such as Americans, demonstrate a higher propensity to switch. Consequently, alternatives, as enablers of switching, were assumed to have a stronger negative impact on loyalty in the USA than in Germany ($H_{C8a\text{-}c}$).

The data collected for this study, however, uncovers that the effect of alternatives on repurchase intentions in Germany is exactly the same as the effect of alternatives on customer loyalty in the USA – the only difference being a lower significance level in the USA. In addition, total effects in Germany (-0.37/-0.17/-0.15 for repurchases/additional purchases/referrals) are very similar to the total effect of -0.25 in the USA, such that support is not offered for $H_{C8a\text{-}c}$. Overall, the influence of alternatives on loyalty is relatively low in both countries and, as pointed out before, alternatives' major effect on loyalty is mediated through satisfaction and proactive improvement, where alternatives serve as a level of comparison.

7.3.7 Overview of cultural differences

As a result of the high dissimilarity of the US and German models, quantitative hypothesis-testing could not be conducted in the preceding sections. Also, of the nine proposed determinants of customer loyalty, only five could be examined regarding potential cultural differences, because fairness, organizational trust, and commitment were excluded from further analyses as a result of lacking discriminance in Germany and/or the USA.

The five determinants that were scrutinized in the preceding sections, service quality, price satisfaction, relational satisfaction, personal trust, and alternatives, only offer limited support of the general assumption that cultural differences exist between Germany and the USA that would have an influence on the formation of customer loyalty. Qualitative support for the respective hypotheses could only be provided for service quality and personal trust. As expected, service quality appears to be more important in loyalty considerations in the USA, as reflected in direct effects and to a limited extent in total effects. In addition, personal trust was excluded from the US customer loyalty model as a result of initial model refitting. This supports the surmise that trust would only play a subordinate role in US 3PL relationships.

The remaining three determinants (price satisfaction, relational satisfaction, and alternatives) do not appear to be different between the USA and Germany and support for the respective hypotheses was therefore not found. The results of the argumentation regarding cultural differences are summarized in Table 7-38.

Table 7-38. Overview of qualitative analysis of cultural differences

Hypothesis	Loyalty Determinant	In the USA compared to Germany	Qualitative Analysis
$H_{C1a\text{-}c}$	Service quality	stronger	some support
$H_{C2a\text{-}c}$	Price satisfaction	stronger	no support
$H_{C3a\text{-}c}$	Relational satisfaction	weaker	no support
$H_{C6a\text{-}c}$	Personal trust	weaker	some support
$H_{C8a\text{-}c}$	Alternatives	stronger	no support

8 Conclusion

After having discussed the results from the empirical studies in the USA and Germany in the preceding chapter, the present chapter will summarize the main findings, present managerial implications, and give recommendations for future research.

8.1 Summary

A thorough review of research in the fields of marketing and logistics revealed that LSPs are confronted with diverse management challenges that result from continuous growth, globalization, and customer diversity. It was also shown that one way to face these challenges is to install an effective and efficient management of customer loyalty. This study therefore aimed at identifying determinants of customer loyalty in relationships between LSPs and their customers, accounting specifically for potential moderating effects of relationship characteristics and cultural context.

In order to tackle the identified research needs, a multi-step approach was chosen. In a first step, social exchange theory, equity theory, and commitment trust theory were drawn on to conceptualize nine determinants of customer loyalty in the context of the relationship between LSPs and their customers. Subsequently, hypotheses were derived to devise a model that includes the identified loyalty determinants, their effects on the three dimensions of customer loyalty, as well as interdependencies between the determinants. Then, four important relationship characteristics were derived from existing research and hypotheses regarding their moderating effects on the direct linkages between customer loyalty and its nine determinants were proposed. Finally, three cultural frameworks were analyzed to detect particular cultural differences between the USA and Germany and as a result, hypotheses on differences between the two countries with respect to the constitution of customer loyalty were formulated.

In a second step, two large-scale empirical studies were conducted. In Germany, a total of 3,402 questionnaires were sent out electronically, of which 17.02% or 579 were completed. After correcting for missing values, a total of 545 responses was used for later analyses. In the USA, 1,448 questionnaires were mailed electronically. Out of 263 responses (18.20%), 250 were usable for analyses. On this basis, measurement models of the previously conceptualized constructs were assessed and structural equation modeling was applied to test the proposed hypotheses. In doing so, the four research questions (RQ) that guided the present study could be answered and findings will be elaborated in the following paragraphs.

RQ_1: What are the relevant determinants of customer loyalty regarding repurchases, additional purchases, and referrals?

In order to identify relevant determinants of customer loyalty, a thorough review of marketing and logistics literature was conducted. Overall, the approach chosen by Wallenburg (2004) proved most appropriate also for this study and was therefore adopted in a modified form. As a result, nine factors were surmised to be of particular relevance in the context of relationships between LSPs and their customers. These are service quality, price satisfaction, relational satisfaction, proactive improvement, alternatives, fairness, affective commitment, personal trust, organizational trust.

Drawing on social exchange theory, equity theory, commitment trust theory, and previous research by Wallenburg (2004), direct effects on the three dimensions of customer loyalty were hypothesized for each of the determinants. In order to validate the effects empirically, however, research question two (RQ2) has to be treated first, as the selected methodology (structural equation modeling) requires that all relationships within the models are specified ex ante.

RQ_2: How can the identified factors be integrated into a comprehensive model of customer loyalty and what interdependencies are observed?

Building on the same theoretical fundament as stated above, relationships between the nine customer loyalty determinants were hypothesized. In this way, 23 relationships were specified (see Figure 4-3). Together with the 27 direct effects on the three dimensions of customer loyalty, a comprehensive explanatory model of customer loyalty was formulated and subjected to empirical validation using the data collected in the USA and Germany.

In Germany, the analysis of discriminant validity using the Fornell/Larcker criterion showed that fairness and organizational trust are not discriminant from the remaining seven loyalty determinants, wherefore these two constructs had to be excluded from further analyses. The resulting seven-factor model of customer loyalty was estimated using the German data set of 545 responses, and as requirements of model fit were fulfilled, the ex ante specified structure could not be rejected. In the following paragraphs, findings regarding the effects on the three loyalty dimensions will be presented.

With an R^2 of over 70%, the construct of repurchase intentions is very well explained by the seven determinants of customer loyalty. In addition, all relationships between the seven determinants and repurchases are significant and positive (with the exception of the direct effect of alternatives on repurchase intentions, which is negative), such that none of the hypotheses can be rejected. With respect to total effects on repurchases, particularly strong relationships are observed for service quality and proactive improvement, which shows that the provision of good service quality and proactive improvement efforts are the keys to customer retention. In both cases, direct effects are limited and the majority of the effects is mediated through other constructs.

Even though relational satisfaction, commitment, and personal trust have significant effects on repurchases, their effects' magnitude is considerably smaller than those of service quality and proactive improvement. While relational satisfaction is a strong driver of commitment and trust, its direct effect on repurchases is relatively weak. Similarly, personal trust has a strong effect on commitment, but only a small direct influence on repurchases. Since commitment does not have other direct effects than those on customer loyalty, its impact on loyalty is not mediated through other constructs. It is therefore interesting to note that its direct effect on repurchases is stronger than all other direct effects and although its total effect is weaker than that of service quality and proactive improvement, commitment can be concluded to play an important role for retention.

Even though cost reductions are readily stated to be the single most important objective in logistics outsourcing decisions, this is neither reflected in direct nor total effects of price satisfaction. The effects on repurchases are negligibly small, indicating that prices actually do not foster retention. Finally, alternatives have a strong negative total effect on repurchase intentions. It shows, however, that the direct effect of alternatives on repurchases is significant but very weak and that the majority of the effect is mediated through proactive improvement and the three satisfaction constructs. While alternatives therefore do not prove to directly affect repurchase considerations, the availability of adequate market alternatives to the currently used LSP does serve as a comparison standard in the evaluation of other factors. It may therefore be concluded that creating retention is the more difficult, the more alternative LSPs are available in the market to carry out a specific logistics activity.

The construct of additional purchase intentions, which grasps a customer's intention to purchase additional logistics services from its currently most important LSP, is only partially explained by the proposed customer loyalty model, as expressed through an R^2 of 32.9%. This may be explained through the fact that the incorporated loyalty determinants are specifically geared at explaining loyalty for services that are already outsourced. With regard to additional purchase considerations, however, an initial outsourcing decision has to be taken, in which strategic issues that are not reflected in the proposed loyalty determinants can play a decisive role.

Examining effects of the individual determinants on additional purchases, commitment and proactive improvement have the greatest explanatory power. Since both constructs are reflections of cognitive and affective processes that need time to develop and unfold their beneficial effects, long-term considerations appear to have a stronger influence on additional purchase deliberations than do shorter-term effects. While service quality and trust still have a distinct effect on additional purchase intentions, price satisfaction's and relational satisfaction's effects are negligible. Similar to the results for repurchases, alternatives have an effect only as a comparison standard in the evaluation of other factors and do not directly influence additional purchases.

The construct of referrals, which refers to a customer's behavior in recommending its most important LSP, is well explained by the seven determinants contained in the model with an R^2 of 54.3%. Examining effects of the determinants individually, commitment and proactive improvement show to be key in creating referrals, as these two constructs not only have the strongest total effects on referrals, but also the highest direct effects. While commitment only has a direct effect on referrals, proactive improvement's direct and total effects are both strong. This indicates that the long-term signaling effect emitted by proactive improvement is just as important for encouraging referral behavior as the immediate performance increasing effect. Even though not quite as strong as the former two effects, service quality is an important factor in stimulating referral behavior. An examination of total and direct effects reveals that displays of good performance increase a customer's propensity to recommend its LSP both by directly affecting referrals, and by influencing other loyalty determinants.

The remaining two satisfaction constructs, relational satisfaction and price satisfaction, as well as personal trust do not appear to be consequential for determining referral behavior. Similarly, alternatives do not directly influence referrals, indicating that market alternatives to the currently most important LSP do not factor into the decision to recommend it to others. However, alternatives have a substantial total effect on referral behavior which is a result of the use of alternatives as a comparison standard in the evaluation of satisfaction and proactive improvement.

Overall, commitment, proactive improvement, and service quality display the strongest effects on customer loyalty in the German model. Apparently, loyalty is not only created as a result of the provision of good performance. At least as important are longer-term factors that assure a customer that the LSP will continue to deliver satisfactory performance and that it will act in the best interest of its customer. In this way, the present study supports the findings obtained by Wallenburg (2004, p. 265), who showed that superior performance together with proactive improvement and a well functioning relationship are the main drivers of customer loyalty.

In the USA, measurement model assessment revealed several deficits. First, the three-dimensional structure of customer loyalty was not validated. Accordingly, a single customer loyalty construct which incorporates aspects of repurchase intentions, additional purchase intentions, and referral behavior was formed. For this reason, the proposed research questions cannot be addressed as formulated and the discussion had to be limited to effects on a comprehensive customer loyalty construct. In addition, fairness and commitment had to be excluded from further analyses, because they are not discriminant from the other constructs. An initial evaluation of the structural model then lead to the elimination of personal trust, organizational trust, and price satisfaction. These three constructs only displayed minute effects on customer loyalty and did not increase the model's explanatory power. Hence, the validated US customer loyalty model only includes the comprehensive customer loyalty construct as well as alternatives, proactive improvement, service quality, and relational satisfaction, which accomplish a very good explanatory power of loyalty with an R^2 of 71.3%.

Of the four loyalty determinants in the model, service quality and proactive improvement turn out to have the greatest effects on customer loyalty. While the major share of the effect of service quality directly influences loyalty, proactive improvement mainly increases loyalty through mediation by service quality. As such, the performance increasing component of proactive improvement clearly outweighs the longer-term oriented signaling effect, which would have been manifested in a strong direct effect on loyalty.

While the remaining two constructs, i.e. relational satisfaction and alternatives, directly affect customer loyalty, their effects are considerably weaker. In fact, alternatives only contribute to the explanation of customer loyalty as a comparison standard for the evaluation of service quality and proactive improvement.

RQ3: Which relationship characteristics are expected to moderate the formation of customer loyalty between LSPs and their customers and which moderating effects on the model of customer loyalty can be observed?

Based on literature review, four relationship characteristics (opportunism, relationship age, centralization of logistics decisions, outsourcing focus) were identified and hypotheses were derived regarding their moderating effects on the model of customer loyalty. Empirical analyses showed, however, that the model is moderated in only very few cases, both in the USA and in Germany.

In Germany, a total of 72 hypotheses were tested, of which 64 had to be rejected. Thus, moderations can only be expected in less than 12% of cases. While no moderations are supported for the degree of centralization of logistics decisions, some scattered moderations are asserted for the three remaining moderators, i.e. opportunism, relationship age, and outsourcing focus. However, none of the moderations appear systematic and their conclusive power is therefore limited.

In the USA, 15 moderation hypotheses were tested in total, of which only three could not be rejected (20%). These three hypotheses refer to opportunism and relationship age, while no moderations were found for the centralization of logistics decisions and the outsourcing focus. Similar to Germany, however, the identified moderations do not possess conclusive power, as they are not systematic.

Overall, both the German and the US model are only scarcely moderated by the four proposed relationship conditions. Rather than suggesting that customer diversity indeed influences the formation of customer loyalty, the results imply that the proposed customer loyalty model is robust against contingency factors not explicitly included in the model. As such, the general applicability of the three theories used for model development is underlined. Also, results are in line with findings obtained by Wallenburg (2004, p. 264), who showed that neither the logistics development stage, nor the strategic importance of logistics moderate his customer loyalty model.

RQ4: What differences between the USA and Germany regarding the model of customer loyalty are expected and is there empirical evidence suggesting that these differences actually exist?

On the grounds of the cultural frameworks developed by Hofstede (2001), Trompenaars and Hampden-Turner (1997), and Hall and Hall (1990), ten cultural differences between the USA and Germany were identified and applied to the customer loyalty model. As a result, eight moderation hypotheses were derived that predicted differences between the two countries with regard to the loyalty effects of service quality, relational satisfaction, price satisfaction, fairness, commitment, personal trust, organizational trust, and alternatives.

However, as the customer loyalty models in Germany and the USA are structurally highly dissimilar, the proposed hypotheses could not be tested statistically. Firstly, customer loyalty itself was found to differ between the two countries. While its three dimensional structure is supported by the German data, model assessment in the USA yielded a uni-dimensional customer loyalty construct. This may be attributable to differing notions of loyalty, a cultural bias in measurement model development, and/or the rigorous model assessment criteria applied in this study. Secondly, the models are distinct with regard to the customer loyalty determinants contained. In Germany, the customer loyalty model comprises seven determinants, while only four are included in the US model.

Since quantitative hypothesis-testing was therefore impossible, the proposed moderations could only be evaluated qualitatively by comparing relationship strengths in Germany and the USA. However, due to the structural differences between the models, only five out of the eight relationships for which moderation hypotheses had been formulated could be discussed. Of these, no indication of any difference was found for the effects of the following three constructs on customer loyalty. Contrary to the assumed focus of US-Americans on short-term economic benefits, price satisfaction is not more important in the USA than in Germany. In fact, price satisfaction was eliminated from the US model due to a negligibly weak impact on loyalty. Similarly, alternatives are not more important in the USA than in Germany, even though US-Americans were surmised to display a higher propensity to switch. Finally, relational satisfaction was expected to have a stronger loyalty effect in Germany, where it is generally presumed that relationships are more important than in the USA. Nevertheless, the empirical data suggests that the effect of relational satisfaction on loyalty is equally strong in both countries.

In contrast to that, two hypotheses find some empirical support. Service quality has a stronger effect in the USA than in Germany. This is in line with previous findings of other authors who concluded that US-Americans are more outcome-focused than Germans, wherefore service quality is more crucial for relationship maintenance in the USA. Also, personal trust shows to be different in the two countries. The respective hypothesis attributed a lower importance to trust in the USA than in Germany, as trust was not expected to have a strong effect in the USA, where interpersonal relationships in the business world are not as important as in Germany. This argumentation finds some support, even though the only substantial loyalty effect of personal trust in Germany is geared at repurchase intentions. In the US, however, personal trust was excluded from the model, as it was found to only have a very small effect on customer loyalty.

Overall, Germans and US-Americans appear to display more similarities than differences. This is not surprising, though, since the three cultural frameworks included in the considerations suggested only subtle differences between the two countries. Nevertheless, some differences could be identified with regard to the relevance of trust and service quality in loyalty considerations. While service quality is very important in both countries, its impact is even more pronounced in the USA. In contrast to that, personal trust, which was discovered to merely play a subordinate role in German customer loyalty considerations, is not even part of the US customer loyalty model, tending to its minute effect on loyalty.

8.2 Managerial implications

The findings from the present research have various managerial implications, which will be briefly presented in the following:

Focus on service quality, proactive improvement, and commitment.
Apart from service quality, which is clearly an important factor in any logistics outsourcing relationship, proactive improvement efforts and the creation of commitment within the relationship are important drivers of customer loyalty. Proactive improvement does not only lead to short-term performance improvements, when improvements are actually implemented, but also sends a positive signal that the LSP continuously pursues proactive improvements in the best interest of its customer. Similarly, a committed customer is an effective means of assuring a relationship's long-term orientation, which is at the core of any customer loyalty considerations. In order to create commitment, customers must be satisfied with the interaction in the relationship and have to trust their LSP.

Price only has a limited effect on customer loyalty.
Even though cost reductions are an important factor when taking the initial decision to outsource logistics services, satisfaction with prices only has a small impact on loyalty considerations. While this should not suggest that customers do not pay any attention to price once engaged in an outsourcing relationship, this study indicates that price is not an order winning criterion, but merely an order qualifier. It can therefore be expected that offering low prices does not increase customer loyalty, while offering high prices could actually be detrimental.

Customer diversity is not as important as usually assumed.
Although most LSPs are serving a truly diverse clientele, this study suggests that differences between customers do not substantially influence the constitution of customer loyalty. Rather, customer loyalty appears to be a result of primarily subjective evaluations of the performance of and the relationship with an LSP, which should be shared by most customers, independent from the relational environments they act in. Nevertheless, customer attributes should be considered carefully when devising concrete efforts for increasing customer loyalty.

Cultural differences have to be observed when doing business internationally.
This study found some evidence that the mechanisms underlying the formation of customer loyalty differ depending on the national context customers are settled in. While the study does not conclusively determine which factors to stress in the USA and Germany, a thorough preparation and familiarization with cultural peculiarities can be advised whenever outsourcing arrangements cross national boundaries.

8.3 Recommendations for further research

In addition to having managerial implications, this study also provides several suggestions for further research, which will be addressed in the following paragraphs:

(1) While the customer loyalty model validated in this study possesses good power for explaining repurchase intentions and referral behavior, only partial explanation of the construct of additional purchase intentions is achieved. As stated before, factors not contained in the model such as strategic outsourcing considerations can be assumed to affect the intention of customers to outsource additional logistics activities to the currently most important LSP. For this reason, future studies should explore additional determinants of this loyalty dimension.

(2) Measurement model assessment revealed that the operationalization of fairness in this study does not achieve sufficient discriminance from other constructs, especially from trust and relational satisfaction. As there is a strong theoretical indication that fairness is important in customer loyalty considerations, further studies should modify fairness' measurement model, e.g. by more strongly recurring to the concept of inequity.

(3) Measurement model assessment also showed that the ex ante surmised three dimensional structure of customer loyalty is not supported by the empirical data collected in the USA. Three possible reasons for this were identified and should be thoroughly considered in further studies. A suitable starting point for this would be to conduct a qualitative pre-study in the US to explore whether the three-dimensional conceptualization matches the notion of customer loyalty in US-Americans.

(4) This study, as well as studies conducted by Wallenburg (2004), Engelbrecht (2004), and Deepen (2006), concludes that proactive improvement is an important factor in relationships between LSPs and their customers. Therefore, further research efforts should elaborate this construct. On the one hand, antecedents of proactive improvements should be investigated. On the other hand, this study suggests that proactive improvement has an immediate performance-increasing effect and a long-term signaling effect that contributes to building an LSP's reputation as a customer-focused and proactive service provider. In order to provide further insights into the exact way the effects of proactive improvement unfold, differentiated outcome components should be identified and empirically examined.

(5) Within this study, four relational characteristics were examined. In addition, analyses were conducted for a multitude of other contingency factors that are not included in the present study. Overall, however, no conclusive moderations were identified. Nevertheless, it may be assumed that customer diversity still has moderating effects on the formation of customer loyalty. The determinants contained in this study, however, capture rather general evaluations of relationships between LSPs and their customers, which may be too broad to be subject to moderating effects. For this reason it would be sensible to examine antecedents of the employed determinants, as moderating effects could surface when this level of detail is added to the analyses.

(6) Within this study, customer loyalty models were validated for the USA and Germany which exhibit considerable structural differences, wherefore hypothesis testing could not be conducted. In order to be able to apply structural equation modeling to comparative intercultural research in future studies, aspects of culture, e.g. on the basis of Hofstede's (2001) five dimensions of culture, should be operationalized and included in surveys along with the items devised for the structural models. In this manner, preliminary analyses could be conducted to validate that cultural differences actually exist between two data sets, i.e. if the cultural items are significantly different. If they are, such items

could be used as moderators to test for differences even between two structurally dissimilar models.

(7) Finally, further research should be conducted employing comparative intercultural research in the field of logistics and customer loyalty. However, as the results of this study indicate that only subtle differences exist between Germany and the USA, further research should explore different cultural settings, e.g. by comparing European countries or the USA with Asian countries.

Appendix A: Questionnaire (Germany)

WHU-BVL-Studie:
"Erfolg durch Logistik-Outsourcing"

Prof. Dr. Jürgen Weber
Dipl.-Kfm. David L. Cahill
Dipl.-Kfm. Jan Deepen

Kühne-Zentrum für Logistikmanagement
WHU - Otto-Beisheim-Hochschule
Burgplatz 2, 56179 Vallendar

Als Dankeschön für Ihre Teilnahme erhalten Sie:

- **einen individuellen Benchmarking-Bericht** (Vergleich Ihres Unternehmens mit den Werten Ihrer Branche, der Ihnen nach Abschluss der Studie zugesandt wird.)

  sowie zusätzlich

- **eine Ausgabe des Buches** "E-Commerce in der Logistik: Quantensprung oder business as usual?" (151 S.) von Prof. Dr. Jürgen Weber et al.

  oder

- **die kostenfreie Teilnahme am 3. WHU Logistiksymposium** (2005) an der WHU

Bitte lesen Sie die folgenden Hinweise, bevor Sie den Fragebogen ausfüllen:

- Alle Daten werden **anonym** ausgewertet und **streng vertraulich** behandelt

- **Bitte füllen Sie alle Fragen so gut wie möglich aus**, auch wenn manche Fragen ähnlich erscheinen. Dies ist aus methodischen Gründen nötig. Wenn Sie die genaue Antwort nicht kennen, bitten wir Sie bewußt um Ihre subjektive Einschätzung.

- **Bitte beziehen Sie alle Fragen zu Ihrem Unternehmen immer auf Ihre Geschäftseinheit** bzw. auf den Teilbereich des Unternehmens, für dessen Logistik Sie (mit)verantwortlich sind.

- Falls Sie verschiedene Logistikdienstleister (LDL) benutzen, **beziehen Sie alle Fragen auf den aus Ihrer Sicht für Ihr Unternehmen bzw. Ihre Geschäftseinheit wichtigsten LDL**.

- Der in dieser Studie verwendete Begriff **Logistik-Outsourcing bezieht sich auf die dauerhafte Fremdvergabe logistischer Dienstleistungen an einen Logistikdienstleister**.

- **Sie können die Beantwortung des Fragebogens nach jeder abgeschickten Seite unterbrechen** und dann durch den in der Email enthaltenen Link fortsetzen. Die eingegebenen Daten werden automatisch nach jeder Seite gespeichert. **Das Dankeschön gibt es für alle vollständig ausgefüllten Fragebögen**.

- Bitte verwenden Sie zur Navigation innerhalb des Fragebogens **ausschließlich die "zurück" und "weiter" Felder**, da es sonst zu technischen Problemen kommen kann.

Für Fragen stehen Ihnen Dipl.-Kfm. David Cahill oder Dipl.-Kfm. Jan Deepen gerne unter 0261-6509-471 oder cahill@whu.edu zur Verfügung!

A. Allgemeine Fragen zum Logistik-Outsourcing

Ein wesentlicher Grund für uns, einen Teil unserer Logistik outzusourcen, war/waren:	*Trifft gar nicht zu*		*Trifft voll zu*
Die Senkung der Logistikkosten.	O O O O O O O		
Die Variabilisierung unserer Fixkosten.	O O O O O O O		
Der Spitzenausgleich bei Kapazitätsschwankungen.	O O O O O O O		
Die niedrigere Kapitalbindung.	O O O O O O O		
Der Geschwindigkeitsgewinn für unsere Logistik.	O O O O O O O		
Die flexibleren Abläufe und kürzeren Reaktionszeiten.	O O O O O O O		
Die geringere Schadens- und Fehlerquote.	O O O O O O O		
Die verbesserte Lieferfähigkeit.	O O O O O O O		
Die Tatsache, dass dieser LDL ein deutlich höheres Logistik--Know-how besitzt.	O O O O O O O		
Die Tatsache, dass Logistik für uns zu den eher unwichtigen Prozessen gehört.	O O O O O O O		
Der Engpass bei unseren Managementkapazitäten.	O O O O O O O		

Wichtig: Bitte beziehen Sie sich ab hier immer auf den für Sie wichtigsten Logistikdienstleister!

	Grad der Durchführung			*Leistungs-schwerpunkt*		
Welche Logistikleistungen führt der für Sie wichtigste Logistikdienstleister (LDL) mit welchem Schwerpunkt aus?	*Gar nicht*		*Voll-ständig*	*Be-schaf-fung*	*Pro-duk-tion*	*Dis-tribu-tion*
Planung und Steuerung der Transporte	O O O O O O O			O	O	O
Transportdurchführung	O O O O O O O			O	O	O
Internationales Freight Forwarding (Frachtversand)	O O O O O O O			O	O	O
Zollabfertigung	O O O O O O O			O	O	O
Cross-Docking	O O O O O O O			O	O	O
Lagerhaltung	O O O O O O O			O	O	O
Lager-Management	O O O O O O O			O	O	O
Verpacken, Kommissionieren	O O O O O O O			O	O	O
Montagetätigkeiten	O O O O O O O			O	O	O
Retourenabwicklung	O O O O O O O			O	O	O
IT-Systeme der Logistik	O O O O O O O			O	O	O
Logistik-Koordination (Lead Logistics Management)	O O O O O O O			O	O	O
Beratungsleistungen	O O O O O O O			O	O	O

B. Zusammenarbeit mit Ihrem Logistikdienstleister

Inwieweit treffen die folgenden Aussagen bzgl. Ihrer Zufriedenheit über die Zusammenarbeit mit Ihrem Logistikdienstleister (LDL) zu?	*Trifft gar nicht zu*		*Trifft voll zu*
Die Ziele und Erwartungen, die wir im Vorfeld der Zusammenarbeit festgelegt hatten, werden durch diesen LDL vollständig erfüllt.	O O O O O O O		
Wir sind sehr zufrieden mit diesem LDL.	O O O O O O O		
Die Zusammenarbeit mit diesem LDL kann man als sehr gut bezeichnen.	O O O O O O O		
Dieser LDL erbringt seine Leistungen in der geforderten Qualität.	O O O O O O O		
Dieser LDL erbringt seine Leistungen in der geforderten Zeit.	O O O O O O O		
Unsere Logistikkosten sind durch das Outsourcing in dem von uns vorhergesagten Ausmaß gesenkt worden.	O O O O O O O		
Mit der Art und Weise des Umgangs sind wir sehr zufrieden.	O O O O O O O		
Konflikte in der Zusammenarbeit mit diesem LDL werden immer reibungslos beigelegt.	O O O O O O O		
Die Beziehung zu diesem LDL kann man als sehr gut bezeichnen.	O O O O O O O		
Dieser LDL gibt eigene Einsparungen aus verbesserten Prozessabläufen etc. in angemessenem Umfang an uns weiter.	O O O O O O O		
Dieser LDL bereichert sich auf unsere Kosten.	O O O O O O O		
Wir fühlen uns durch diesen LDL fair behandelt.	O O O O O O O		
Wir profitieren in gleichem Maße vom Outsourcing wie dieser LDL.	O O O O O O O		

Inwieweit stimmen Sie den folgenden Aussagen hinsichtlich Ihres Verhältnisses zu Ihrem Logistikdienstleister zu?	*Trifft gar nicht zu*		*Trifft voll zu*
Dieser LDL hält immer die Zusagen, die er uns macht.	O O O O O O O		
Dieser LDL ist bei auftretenden Problemen immer offen und ehrlich zu uns.	O O O O O O O		
Dieser LDL ist sehr vertrauenswürdig.	O O O O O O O		
Dieser LDL ist stark daran interessiert, dass wir erfolgreich sind.	O O O O O O O		
Auch Dinge, die wir nicht oder nur unter großem Aufwand kontrollieren können, erledigt dieser LDL	O O O O O O O		
Bei wichtigen Entscheidungen berücksichtigt der LDL auch unsere Interessen.	O O O O O O O		
Wir verteidigen diesen LDL, wenn er durch Mitglieder unseres Unternehmens oder durch Personen von	O O O O O O O		
Wir würden es persönlich sehr bedauern, wenn wir die Geschäftsbeziehung mit diesem LDL aufgeben	O O O O O O O		
Wir fühlen uns persönlich angegriffen, wenn dieser LDL durch Mitglieder unseres Unternehmens oder durch	O O O O O O O		
Wir haben den festen Willen, die Beziehung so lange wie möglich aufrechtzuerhalten.	O O O O O O O		

Inwieweit stimmen Sie den folgenden Aussagen zur Kooperation mit diesem Logistikdienstleister zu?	*Trifft gar nicht zu*						*Trifft voll zu*
Die Ziele unserer Zusammenarbeit wurden von uns und diesem LDL gemeinsam festgelegt.	O	O	O	O	O	O	O
Hinsichtlich der Art und Weise, zusammen zu arbeiten, Geschäfte zu machen und Projekte durchzuführen,	O	O	O	O	O	O	O
Dieser LDL und wir ziehen in allen Belangen an einem Strang.	O	O	O	O	O	O	O
Wenn auf übergeordneter Ebene noch Probleme oder Unklarheiten hinsichtlich der Outsourcing-Kooperation auftauchen, fällen wir notwendigen Entscheidungen gemeinsam mit dem LDL.	O	O	O	O	O	O	O
Wenn im Rahmen unserer Geschäftsbeziehung einer der Partner seine Macht ausspielt, geschieht das in einer angemessenen Art und Weise.	O	O	O	O	O	O	O
In unserer Geschäftsbeziehung respektieren sich beide Seiten vollkommen.	O	O	O	O	O	O	O
Auch abseits der vorher festgelegten Zuständigkeiten arbeiten unsere Mitarbeiter mit dem LDL zusammen, um den Erfolg der Kooperation sicherzustellen.	O	O	O	O	O	O	O
Die Kooperation mit diesem LDL funktioniert sehr gut.	O	O	O	O	O	O	O

Inwieweit beschreiben die folgenden Aussagen die relevanten Teilschritte auf dem Weg zum Outsourcing?	*Trifft gar nicht zu*						*Trifft voll zu*
Einige Führungskräfte in unserem Unternehmen sind gegen die Outsourcing-Entscheidung.	O	O	O	O	O	O	O
Auf Seiten der operativen Mitarbeiter in unserem Unternehmen gibt es viele Widerstände gegen das Outsourcing-Projekt.	O	O	O	O	O	O	O
Zwischen dem LDL und uns treten immer wieder Konflikte auf der Managementebene auf.	O	O	O	O	O	O	O
Auf der operativen Ebene kommt es häufig zu Problem oder Konflikten zwischen unseren Mitarbeitern und denen des LDL.	O	O	O	O	O	O	O
Insgesamt ist die Beziehung zu diesem LDL sehr konfliktreich.	O	O	O	O	O	O	O

Inwieweit beschreiben die folgenden Aussagen die laufende Zusammenarbeit mit diesem Logistikdienstleister?	*Trifft gar nicht zu*		*Trifft voll zu*
Wir besprechen mögliche Probleme und Verbesserungen regelmäßig mit den verantwortlichen Mitarbeitern des LDL.	O O O O O O O		
Der Austausch von Informationen zwischen unseren Mitarbeitern und denen dieses LDL verläuft sehr gut.	O O O O O O O		
Um unsere gesteckten Ziele zu erreichen, sind viele Treffen und Gespräche mit diesem LDL notwendig.	O O O O O O O		
Wenn wir Informationen mit diesem LDL austauschen, sind diese immer relevant für das Vorankommen des Projektes bzw. unserer Geschäftsbeziehung.	O O O O O O O		
Der Austausch von Informationen zwischen diesem LDL und uns findet immer sofort statt, sobald sie zur Verfügung stehen.	O O O O O O O		
Auf die Informationen, die zwischen uns und diesem LDL ausgetauscht werden, können sich beide Seiten vollkommen verlassen.	O O O O O O O		
Die Art und Weise, wie wir mit diesem LDL Informationen austauschen, ist besonders gut dazu geeignet, Probleme in beiderseitigem Interesse zu lösen.	O O O O O O O		

Inwieweit stimmen Sie den folgenden Aussagen zu dem Verhältnis bzw. der Zusammenarbeit mit diesem Logistikdienstleister zu?	*Trifft gar nicht zu*		*Trifft voll zu*
Der LDL war frühzeitig und umfassend eingebunden.	O O O O O O O		
Für das mit diesem LDL durchgeführte Outsourcing ist ein funktionsübergreifendes Projektteam verantwortlich, in dem alle relevanten Funktionen unseres Unternehmens vertreten sind.	O O O O O O O		
Die auf beiden Seiten für das Outsourcing verantwortlichen Mitarbeiter arbeiten sehr gut zusammen.	O O O O O O O		
Den alltäglichen Austausch zwischen uns und diesem LDL kann man als informell bezeichnen.	O O O O O O O		
Wenn Fragen oder Probleme auftreten, können wir diesen LDL spontan kontaktieren und sofort gemeinsam beginnen, an Lösungen zu arbeiten.	O O O O O O O		
Die Beziehung zwischen uns und diesem LDL ist sehr offen.	O O O O O O O		
Wir tauschen mit diesem LDL auch sehr sensitive Daten aus, wenn wir uns dadurch einen Vorteil erhoffen.	O O O O O O O		
Wenn eine der beiden Parteien mit etwas nicht zufrieden ist, sagen wir uns das klar und deutlich.	O O O O O O O		
Der LDL ist immer vollkommen offen und ehrlich zu uns.	O O O O O O O		

Inwieweit stimmen Sie den folgenden Aussagen zu Ihrem Verhältnis bzw. der Zusammenarbeit mit diesem Logistikdienstleister zu?	*Trifft gar nicht zu*						*Trifft voll zu*
Dieser LDL und wir bewältigen auftretende Problem immer gemeinsam.	O	O	O	O	O	O	O
Der Informationsaustausch bei der Behandlung von Problemen funktioniert sehr gut.	O	O	O	O	O	O	O
Wenn Probleme in der Zusammenarbeit mit diesem LDL auftauchen, werden sie immer von derselben der beiden Parteien angefasst und beseitigt.	O	O	O	O	O	O	O
Wenn es zwischen dem LDL und uns zu Problemen kommt, wird die Diskussion oft unsachlich und auch durch den Austausch von "Unfreundlichkeiten" geprägt.	O	O	O	O	O	O	O
Probleme zwischen dem LDL und uns haben in der Vergangenheit die Produktivität unserer Beziehung stark beeinträchtigt.	O	O	O	O	O	O	O
Probleme zwischen dem LDL und uns werden als Chance verstanden, die beiden Partnern die Möglichkeit zu Verbesserungen eröffnen.	O	O	O	O	O	O	O

Inwieweit treffen die folgenden Aussagen auf die persönlichen Beziehungen zwischen Mitarbeitern Ihres Unternehmens und denen des Logistikdienstleisters zu?	*Trifft gar nicht zu*						*Trifft voll zu*
Wir pflegen viele persönliche Kontakte mit diesem LDL.	O	O	O	O	O	O	O
Die Zusammenarbeit mit diesem LDL klappt auch auf der persönlichen Ebene sehr gut.	O	O	O	O	O	O	O
Wenn wir die Beziehung zu diesem LDL beenden würden, würden ich oder einige meiner Mitarbeiter und Kollegen einen guten Geschäftsfreund verlieren.	O	O	O	O	O	O	O
Wir haben enge persönliche Beziehungen zu Mitarbeitern dieses LDL.	O	O	O	O	O	O	O

Inwieweit treffen folgende Aussagen auf Ihren Hauptansprechpartner beim Logistikdienstleister zu?	*Trifft gar nicht zu*						*Trifft voll zu*
Diese Person ist sehr vertrauenswürdig.	O	O	O	O	O	O	O
Diese Person macht keine falschen Versprechen.	O	O	O	O	O	O	O
Diese Person besitzt großes Know-how und ist ein guter Manager.	O	O	O	O	O	O	O
Diese Person spricht die gleiche Sprache wie wir.	O	O	O	O	O	O	O
Diese Person berücksichtigt auch unsere Interessen und Bedürfnisse.	O	O	O	O	O	O	O

Welche Position hat Ihr Hauptansprechpartner beim LDL?					
O	Geschäftsführer/Vorstand	O	Niederlassungsleiter	O	Bereichsleiter
O	Vertriebsleiter	O	Key Account Manager	O	Andere: ________________

Inwieweit treffen folgende Aussagen zur Verbesserung Ihrer Logistiksysteme durch diesen Logistikdienstleister zu?	*Trifft gar nicht zu*		*Trifft voll zu*
Der LDL arbeitet intensiv daran, die Logistikprozesse fortlaufend zu optimieren.	O O O O O O O		
Der LDL gibt uns laufend Anstöße zu Verbesserungen auch außerhalb seines direkten Zuständigkeitsbereiches.	O O O O O O O		
Bei veränderten Rahmenbedingungen modifiziert der LDL aus eigenem Antrieb die Logistiksysteme bzw. -abläufe soweit sinnvoll und notwendig.	O O O O O O O		
Der LDL spricht uns aus Eigeninitiative mit Verbesserungsvorschlägen an.	O O O O O O O		
Dieser LDL ist generell sehr innovativ.	O O O O O O O		

Inwieweit treffen folgende generelle Aussagen zu diesem Logistikdienstleister zu?	*Trifft gar nicht zu*		*Trifft voll zu*
Der LDL bietet insgesamt einen exzellenten Service.	O O O O O O O		
Die Leistungen des LDL sind hervorragend.	O O O O O O O		
Der LDL bietet eine sehr hohe Qualität.	O O O O O O O		

Inwieweit treffen die folgenden Fragen bzgl. Ihrer Zufriedenheit über die Zusammenarbeit mit diesem Logistikdienstleister zu?	*Trifft gar nicht zu*		*Trifft voll zu*
Die Ziele und Erwartungen, die wir im Vorfeld der Zusammenarbeit festgelegt hatten, wurden in deutlichem Maße übertroffen.	O O O O O O O		
Unsere Anforderungen an die Qualität der Leistungen dieses LDL sind deutlich positiv übertroffen worden.	O O O O O O O		
Unsere Logistikkosten sind auf Grund der Zusammenarbeit mit diesem LDL deutlich niedriger als erwartet.	O O O O O O O		
Die tatsächlichen Kosten sind im Verhältnis zur erbrachten Gesamtleistung wesentlich besser, als wir es vorher erwartet hätten.	O O O O O O O		

Inwieweit treffen folgende Aussagen zur bisherigen und zukünftigen Zusammenarbeit mit diesem Logistikdienstleister zu?	*Trifft gar nicht zu*		*Trifft voll zu*
Wir werden diesen LDL auch zukünftig weiter nutzen.	O O O O O O O		
Aus heutiger Sicht gehen wir davon aus, vorhandene Verträge mit dem LDL bei deren Auslaufen zu verlängern.	O O O O O O O		
Wenn wir mit unserem heutigen Wissen nochmals vor der ursprünglichen Entscheidung über die Zusammenarbeit mit diesem LDL stünden, würden wir die Geschäftsbeziehung erneut eingehen.	O O O O O O O		
Wir werden die Leistungen, die wir von diesem LDL in Anspruch nehmen, bei Auslaufen des Vertrages höchst wahrscheinlich nicht neu ausschreiben, sondern direkt mit diesem LDL verhandeln.	O O O O O O O		

Inwieweit treffen folgende Aussagen zu Ihrer Einstellung gegenüber diesem Logistikdienstleister zu?	Trifft gar nicht zu		Trifft voll zu
Wir haben in der Organisation angeregt, diesen LDL für zukünftige Projekte bevorzugt zu berücksichtigen.	O O O O O O O		
Ich erwähne diesen LDL gegenüber Kollegen häufig sehr positiv.	O O O O O O O		
Ich empfehle diesen LDL auch nach außen hin häufig weiter.	O O O O O O O		
Wir empfehlen diesen LDL häufig weiter.	O O O O O O O		

Inwieweit treffen folgende Aussagen bzgl. der Investitionen in die Geschäftsbeziehung zu?	Trifft gar nicht zu		Trifft voll zu
Wir haben viel Arbeitszeit (Mannjahre) investiert, die umsonst wäre, wenn wir den LDL wechseln würden.	O O O O O O O		
Wir haben hohe Sachinvestitionen getätigt, die erheblich an Wert verlieren würden, wenn wir den LDL wechselten.	O O O O O O O		
Wir haben hohe Investitionen in IT-Systeme getätigt, die erheblich an Wert verlieren würden, wenn wir den LDL wechselten.	O O O O O O O		
Wir haben sehr viel in die Weiterentwicklung des LDL investiert.	O O O O O O O		
Wir haben viel Arbeitszeit (Mannjahre) in die Integration unserer Prozesse mit denen des LDL investiert, die umsonst wäre, wenn wir den LDL wechseln würden.	O O O O O O O		

Bitte geben Sie den Grad der Übereinsstimmung an, den Sie in den folgenden Fragen zwischen Ihrem Unternehmen und diesem Logistikdienstleister vermuten:	Sehr niedrige Überein-stimmung		Sehr hohe Überein-stimmung
Mitarbeiter, die nur auf den eigenen Vorteil bedacht sind anstatt den Vorteil der Firma zu verfolgen, sollten zurechtgewiesen werden.	O O O O O O O		
In Beziehungen sollten Unternehmen nicht nur auf den eigenen kurzfristigen Vorteil bedacht sein, sondern vielmehr den langfristigen Nutzen für die beteiligten Unternehmen im Auge haben.	O O O O O O O		
Die Mitarbeiter beider Seiten haben die Ziele der Zusammenarbeit vollständig verstanden und handeln auch danach.	O O O O O O O		
Die Ziele der Zusammenarbeit sind vollkommen klar definiert und werden von beiden Unternehmen in gleicher Art und Weise verfolgt und angestrebt.	O O O O O O O		
Das grundlegende Verständnis über die Art der Zusammenarbeit ist bei beiden Seiten sehr ähnlich und kompatibel.	O O O O O O O		

Inwieweit treffen folgende generelle Aussagen zu diesem Logistikdienstleister zu?	Trifft gar nicht zu		Trifft voll zu
Die Preise dieses LDL sind im Vergleich zu anderen LDL sehr günstig.	O O O O O O O		
Das Preis-Leistungs-Verhältnis bei diesem LDL ist sehr gut.	O O O O O O O		
Die Preise dieses LDL sind im Vergleich zur Eigenerstellung sehr günstig.	O O O O O O O		

Inwieweit treffen folgende Aussagen zur bisherigen und zukünftigen Zusammenarbeit mit diesem Logistikdienstleister zu?	*Trifft gar nicht zu*		*Trifft voll zu*
In Zukunft wird dieser LDL einen größeren Anteil an unserem Auftragsvolumen erhalten.	O O O O O O O		
Beim Outsourcing anderer als der bisherigen Logistikleistungen werden wir diesen LDL bevorzugt berücksichtigen.	O O O O O O O		
Neue Leistungen werden wir zunächst diesem LDL anbieten, bevor wir sie ausschreiben.	O O O O O O O		
In den nächsten Jahren werden wir stärker auf diesen LDL zurückgreifen als bisher.	O O O O O O O		

Inwieweit beschreiben die folgenden Aussagen die laufende Zusammenarbeit mit diesem Logistikdienstleister?	*Trifft gar nicht zu*		*Trifft voll zu*
Dieser LDL verändert manchmal Fakten so, dass sich seine Interessen besser begründen und vertreten lassen.	O O O O O O O		
Was der LDL uns gegenüber als seine momentanen Aktivitäten und Leistungen für uns angibt, entspricht immer der Realität.	O O O O O O O		
Um seine eigenen Ziele besser zu erreichen, verspricht uns der LDL manchmal Dinge, die er später überhaupt nicht einhält.	O O O O O O O		
Wenn der LDL mit uns über seine Bedürfnisse spricht, übertreibt er manchmal, um seine Ziele besser zu erreichen.	O O O O O O O		
Dieser LDL würde alles in seiner Macht stehende tun, um seine eigenen Ziele zu erreichen.	O O O O O O O		
Dieser LDL hat das Gefühl, dass sich im Umgang mit uns Ehrlichkeit lohnt.	O O O O O O O		

Inwieweit treffen folgende Aussagen zu Ihrer Einstellung gegenüber diesem Logistikdienstleister zu?	*Trifft gar nicht zu*		*Trifft voll zu*
Meiner Meinung nach haben wir zuviel investiert, um einen Abbruch der Geschäftsbeziehung in Betracht zu ziehen.	O O O O O O O		
Auch wenn wir es wollten, wäre es sehr schwierig, die Geschäftsbeziehung zu diesem LDL aufzugeben.	O O O O O O O		
Wir fühlen uns verpflichtet, die Geschäftsbeziehung zu diesem LDL aufrechtzuerhalten.	O O O O O O O		
Dieser LDL hat zuviel in die Beziehung investiert, als dass wir ihm einen Abbruch der Geschäftsbeziehung zumuten könnten.	O O O O O O O		

Inwieweit treffen folgende Aussagen zu Alternativen und Ihrer Abhängigkeit von diesem Logistikdienstleister zu?	*Trifft gar nicht zu*		*Trifft voll zu*
Es gibt viele alternative LDL, die gleichwertige Leistungen bei gleichen Kosten anbieten können.	O O O O O O O		
Es gibt viele alternative LDL, die gleichwertige Leistungen bei höheren Kosten anbieten können.	O O O O O O O		
Es gibt viele alternative LDL, die gleichwertige Leistungen bei niedrigeren Kosten anbieten können.	O O O O O O O		
Dieser LDL beeinflusst die Qualität und Zuverlässigkeit unserer Logistikprozesse so sehr, dass sein Wegfall uns sehr stark treffen würde.	O O O O O O O		
Bei einem Wechsel des LDL würden insbesondere die neuen Verhandlungen sehr viel Aufwand verursachen.	O O O O O O O		
Die Suche und Auswahl eines gleichwertigen LDL ist sehr aufwendig.	O O O O O O O		
Wir können diesen LDL leicht ersetzen.	O O O O O O O		
Ein kurzfristiger Wechsel dieses LDL wäre mit sehr viel Aufwand verbunden.	O O O O O O O		

Inwieweit beschreiben die folgenden Aussagen die spezifischen Investitionen, die für den Logistikdienstleister in Ihrem Fall notwendig waren?	*Trifft gar nicht zu*		*Trifft voll zu*
Um Logistikdienstleistungen für uns zu erbringen, musste der LDL hohe Investitionen (Transportmittel, Lagerhäuser o.ä.) tätigen, die er ausschließlich für uns verwenden kann.	O O O O O O O		
Der LDL kann die für uns entwickelten Geschäftsabläufe nicht ohne große Veränderungen für einen anderen Kunden verwenden.	O O O O O O O		
Damit der LDL Leistungen für uns erbringen kann, musste er seine Mitarbeiter so schulen, dass sie dieses Wissen bei einem anderen Kunden nicht verwenden könnten.	O O O O O O O		
Die Investitionen, die dieser LDL für unsere Beziehung in Menschen, Prozesse oder Vermögenswerte tätigen musste, werden sich sehr schnell amortisieren.	O O O O O O O		

Achtung: Bitte beantworten Sie die folgenden Fragen in Bezug auf <u>Ihr Unternehmen</u> und <u>Ihre Kunden.</u>

C. Fragen zu Ihrer Unternehmenssituation

Inwieweit beschreiben die folgenden Aussagen die zur Zeit in Ihren Unternehmen vorherrschende Unsicherheit?	Trifft gar nicht zu		Trifft voll zu
Es fällt uns schwer, das künftige Verhalten unserer Kunden präzise vorherzusehen.	O O O O O O O		
Es kommt oft vor, dass sich Kunden uns gegenüber opportunistisch oder unfair verhalten.	O O O O O O O		
In unserer Branche oder der unserer Kunden gibt es aufgrund der Wettbewerbssituation zur Zeit eine spürbare Unsicherheit über die Zukunft.	O O O O O O O		
Es fällt uns schwer einzuschätzen, wie sich die Wünsche und Bedürfnisse unserer Kunden in Zukunft entwickeln werden.	O O O O O O O		

Inwieweit beschreiben die folgenden Aussagen die momentane Situation Ihres Unternehmens?	Trifft gar nicht zu		Trifft voll zu
Unser Absatzvolumen ist von Kunde zu Kunde sehr unterschiedlich.	O O O O O O O		
Das Bestellverhalten unser Kunden ist sehr uneinheitlich und schwankt je nach Auftrag oder Saison beträchtlich.	O O O O O O O		
Verglichen mit unseren Wettbewerbern müssen unsere Produkte an eine hohe Anzahl von Lieferpunkten geliefert werden.	O O O O O O O		
Verglichen mit unseren Wettbewerbern greifen wir auf eine sehr große Anzahl von Logistikdienstleistern zurück.	O O O O O O O		
Der Verdrängungswettbewerb innerhalb unserer Branche ist sehr stark.	O O O O O O O		
In den Branchen unserer Kunden sind starke Konzentrationstendenzen zu beobachten.	O O O O O O O		
Die Bedeutung der Logistik nimmt in den Augen unserer Kunden immer weiter zu.	O O O O O O O		
Eine sehr gute Logistik wird von unseren Kunden zunehmend als wichtig empfunden und ist deswegen für uns ein großer Wettbewerbsvorteil.	O O O O O O O		

Inwieweit beschreiben die folgenden Aussagen das Produktprogramm Ihres Unternehmens?	Trifft gar nicht zu		Trifft voll zu
Wir fertigen ein sehr breites Produktprogramm mit einer Reihe von unterschiedlichen Produkten.	O O O O O O O		
Unsere Produkte sind so wertvoll, dass sie besondere logistische Maßnahmen erfordern.	O O O O O O O		
Unsere Produkte sind so unterschiedlich, dass für sie sehr viele unterschiedliche Transport-, Umschlag- und/oder Lagerprozesse konzipiert werden müssen.	O O O O O O O		
Für unsere Produkte gibt es echte Alternativen am Markt, so dass wir uns über eine erstklassige Logistik positiv herausstellen können.	O O O O O O O		

Inwieweit treffen die folgenden eher grundsätzlichen Aussagen zur Ausgestaltung der Logistik Ihrer Geschäftseinheit zu?	*Trifft gar nicht zu*		*Trifft voll zu*
Entscheidungen zu logistischen Fragen werden in unserem Unternehmen sehr zentral gefällt.	O O O O O O O		
Unsere Geschäftseinheit beherrscht einen reibungslosen, durchgängigen, schnellen und störungsarmen Material- und Informationsfluss.	O O O O O O O		
Unsere Geschäftseinheit wird insgesamt sehr fluss- bzw. prozessorientiert geführt.	O O O O O O O		
Unsere Geschäftseinheit besitzt einen deutlich höheren Grad an Fluss- bzw. Prozessorientierung als unsere Wettbewerber.	O O O O O O O		
Sämtliche Prozesse der Leistungserstellung sind in unserer Geschäftseinheit gut aufeinander abgestimmt.	O O O O O O O		
Es existieren in unserer Geschäftseinheit zahlreiche Einzelinteressen, die der Erreichung der Ziele unserer Geschäftseinheit im Wege stehen.	O O O O O O O		

D. Fragen zu Logistik- und Unternehmenserfolg

Wie schätzen Sie die Logistikkosten Ihrer Geschäftseinheit im Vergleich zum Wettbewerb ein?	*Sehr viel schlechter*		*Sehr viel besser*
Logistikkosten relativ zum Umsatz (inkl. Vergütung des LDL)	O O O O O O O		
Lagerkosten	O O O O O O O		
Transportkosten	O O O O O O O		
Durch Logistikprozesse hervorgerufene IT-Kosten	O O O O O O O		
Personalkosten der Logistik	O O O O O O O		

Wie schätzen Sie die Logistikleistung Ihrer Geschäftseinheit im Vergleich zum Wettbewerb ein?	*Sehr viel schlechter*		*Sehr viel besser*
Durchlaufzeiten	O O O O O O O		
Lieferzeit	O O O O O O O		
Lieferfähigkeit	O O O O O O O		
Liefertreue	O O O O O O O		
Lieferflexibilität und Reaktionszeiten (Zeit, Menge)	O O O O O O O		
Schadens- und Fehlerfreiheit der logistischen Prozesse	O O O O O O O		

Wie schätzen Sie den Markterfolg Ihrer Geschäftseinheit im Vergleich zum Wettbewerb ein?	*Sehr viel schlechter*		*Sehr viel besser*
Kundenzufriedenheit	O O O O O O O		
Kundennutzen	O O O O O O O		
Bindung bestehender Kunden	O O O O O O O		
Gewinnung/Akquisition von Neukunden	O O O O O O O		
Erreichung des angestrebten Wachstums	O O O O O O O		
Erreichung des angestrebten Marktanteils	O O O O O O O		

Wie schätzen Sie die Flexibilität Ihrer Geschäftseinheit im Vergleich zum Wettbewerb ein?	*Sehr viel schlechter*		*Sehr viel besser*
Anpassung der Produkte/Dienstleistungen an neue Kundenbedürfnisse	O O O O O O O		
Reaktion auf neue Entwicklungen am Markt	O O O O O O O		
Nutzung neuer Marktchancen	O O O O O O O		

Wie verhält sich Ihrer Einschätzung nach die Umsatzrendite Ihrer Geschäftseinheit im Vergleich zum Wettbewerb?	*Sehr viel schlechter*		*Sehr viel besser*
Unsere Umsatzrendite war im letzten Geschäftsjahr im Vergleich zu unseren Wettbewerbern...	O O O O O O O		
Unsere Umsatzrendite war im Durchschnitt der letzten drei Geschäftsjahre im Vergleich zu der unserer Wettbewerber...	O O O O O O O		
Die Entwicklung unserer Umsatzrendite war in den letzten drei Jahren im Vergleich zu der unserer Wettbewerber...	O O O O O O O		

Vielen Dank, dass Sie den Fragebogen bis hierhin ausgefüllt haben!

Wir bitten Sie jetzt noch, einige statistische Fragen zu beantworten. Wie der gesamte Fragebogen unterliegen auch diese Fragen strengster Vertraulichkeit!

Bitte beantworten Sie folgende Fragen zu Ihrer Person :					
In welcher Funktion sind Sie tätig?		O	Geschäftsführer	O	Leiter Logistik
O	Leiter Teilbereich der Logistik	O	Mitarbeiter Logistikbereich	O	Andere: ________________
Seit wie vielen Jahren sind Sie in dieser Position tätig?				____ Jahre	
Seit wie vielen Jahren sind Sie in Ihrem Unternehmen tätig?				____ Jahre	

Welcher Branche gehört Ihre Geschäftseinheit an? *(Bitte <u>nur eine Branche</u> ankreuzen.)*					
O	Nahrungs- und Genussmittel	O	Maschinen- und Apparatebau	O	Chemie/ Kunststoff
O	Fahrzeugbau	O	Elektrotechnik/ Feinmechanik/ Optik	O	Handel
O	Konsumgüter	O	andere: ________________	O	Health Care

Wie groß ist das Umsatzvolumen Ihrer Geschäftseinheit derzeit? *(in Mio. € / Jahr)*							
bis 10	O	11-24	O	25-50	O	50-100	O
100-250	O	250-500	O	500-1.000	O	Über 1.000	O

Wie viele Mitarbeiter arbeiten ungefähr für Ihre Geschäftseinheit?	____ Mitarbeiter
Wie viele Logistikdienstleister nutzt Ihre Geschäftseinheit insgesamt?	____ Logistikdienstleister
Welchen Anteil an Ihren gesamten Logistikkosten haben externe Logistikdienstleister ungefähr (in %)?	____ Prozent

Füllen Sie die folgenden Fragen bitte wieder für den gleichen Logistikdienstleister aus, für den Sie zuvor auch schon die Fragen beantwortet haben.

Welchen Anteil am Gesamtvolumen der fremdvergebenen Logistikdienstleistungen Ihrer Geschäftseinheit hat dieser Logistikdienstleister?	____ Prozent
Seit wie vielen Jahren arbeiten Sie mit diesem Logistikdienstleister zusammen?	____ Jahre (z.B. 2,5)
Seit wie vielen Jahren arbeiten Sie mit diesem Logistikdienstleister so zusammen, dass Sie von einer engeren Geschäftsbeziehung sprechen würden?	____ Jahre (z.B. 2,5)
Wie lang ist die Gesamtlaufzeit des aktuellen Vertrages mit diesem Logistikdienstleister?	____ Jahre (z.B. 2,5)
Wie lang ist die Restlaufzeit dieses Vertrages mit diesem Logistikdienstleister?	____ Jahre (z.B. 2,5)

Vielen Dank, dass Sie an unserer Studie teilgenommen haben!

Sie können jetzt das Dankeschön für Ihre Teilnahme auswählen.

Bitte wählen Sie hier das Dankeschön für Ihre Teilnahme aus. Alle Teilnehmer erhalten zusätzlich einen individuellen Benchmarking-Bericht.	
Ich habe als Dankeschön bereits am 2. WHU Logistiksymposium in Vallendar teilgenommen.	O
Ich hätte gerne das Buch: "E-Commerce in der Logistik: Quantensprung oder business as usual?" von Prof. Dr. Jürgen Weber et al..	O
Ich möchte am 3. WHU Logistiksymposium (2005) an der WHU teilnehmen.	O

Bitte füllen Sie die folgenden Felder aus, damit wir Sie bzgl. Ihres Dankeschöns kontaktieren können oder heften Sie eine Vistenkarte an. Bitte geben Sie auf jeden Fall Name und E-Mail-Adresse an. Falls Sie sich für das Buch entscheiden, benötigen wir auch Ihre Anschrift.

Nachname: ______________________

Vorname: ______________________

E-Mail-Adresse: ______________________

Unternehmen: ______________________

Straße: ______________________

Postleitzahl: ______________________

Ort: ______________________

Telefon: ______________________

Falls Sie Anmerkungen oder Kritik haben, können Sie uns das im Folgenden gerne mitteilen:

____

____

____

____

Vielen Dank für Ihre Mitarbeit!

Appendix B: Questionnaire (USA)

FISHER COLLEGE OF BUSINESS

WHU Excellence in Management Education 1984 – 2004

Prof. Thomas J. Goldsby
Prof. A. Michael Knemeyer

Fisher College of Business

The Ohio State University

554 Fisher Hall, 2100 Neil Avenue
Columbus, OH 43210

Prof. Juergen Weber

Kuehne-Center for Logistics Management
WHU – Otto Beiseim Graduate School of Management
Burgplatz 2
56179 Vallendar, Germany

The success of this research is highly dependent on your participation. Thank you in advance for your time and effort:

We would like to thank you for participating and offer you:

- **an individual benchmarking report** comparing your company with the other participants in the research (We have information from over 100 companies to-date).

  plus the following gift package:

- a complimentary copy of the recently published book **Supply Chain Management: Processes, Partnerships, Performance**, edited by Douglas M. Lambert (which retails for $50 on amazon.com)

  and

- an opportunity to win a **20 GB Apple iPod** (retail value $299). Estimated odds of winning: 1 in 50.

Please read the following instructions carefully before beginning the survey:

- All responses are **anonymously** and we guarantee the **strictest confidentiality.**
- **Please complete all questions as well as you can**, even if some questions appear similar. If you do not know the exact answer, please provide your best guess.
- **Please refer all questions about your company to your business unit** or to the part of the company for which you are managing logistics.
- When completing this survey, **please select a specific logistics service provider (LSP) on which to focus when answering the questions**. This LSP can either be your largest provider or the one that handles a particularly important logistics acitivity
- When using the term **"Logistics Outsourcing", we are talking about providers of comprehensive logistics services** (either on an asset or non-asset basis) that are ultimately responsible for the quality of customer service rendered.
- **You can suspend the completion of the survey after each page you have submitted** and continue later using the hyperlink and password that were included in your invitation email. Your entries are saved by clicking on the "submit" button at the end of each page.
- **Please note that gifts are offered only to those respondents who complete the entire survey**.
- **Please use the "submit" and "back" buttons within the survey.** Using the "back" and "next" buttons of your browser may result in data loss.

Please contact **Michael Knemeyer** for questions:

Tel.: (614) 292-2507
Email: knemeyer.4@osu.edu

A. General questions on logistics outsourcing

Major reasons for outsourcing parts of our logistics activities were: Please mark all applicable reasons and indicate to what extent they apply.	*Does not apply*						*Fully applies*
The reduction of our logistics costs.	O	O	O	O	O	O	O
To turn fixed costs into variable costs.	O	O	O	O	O	O	O
To level peaks when order volumes vary.	O	O	O	O	O	O	O
To reduce our capital employed in logistics processes.	O	O	O	O	O	O	O
To increase the speed of our logistics.	O	O	O	O	O	O	O
To increase process flexibility and shorten response times.	O	O	O	O	O	O	O
To lower the damage- or error ratio.	O	O	O	O	O	O	O
To increase our ability to deliver.	O	O	O	O	O	O	O
Our LSP has significantly better logistics skills.	O	O	O	O	O	O	O
We consider logistics to be a rather unimportant process.	O	O	O	O	O	O	O
Our management capacities are limited.	O	O	O	O	O	O	O

Important: From now on, please answer all questions for the specific logistics service provider (LSP) you selected to serve as your focal provider.

Which services does your selected logistics service provider (LSP) provide?	*Degree of Control*							*Focus of Service*		
Please mark all applicable services that this specific LSP provides. In addition, please also indicate if the focus of this service lies on the inbound side, internal, or outbound side of the business.	*Not at all*						*Comple-tely*	*Inbou-nd*	*Inter-nal*	*Out-bou-nd*
Transportation Planning	O	O	O	O	O	O	O	O	O	O
Transportation Operations	O	O	O	O	O	O	O	O	O	O
International Freight Forwarding	O	O	O	O	O	O	O	O	O	O
Customs Clearance	O	O	O	O	O	O	O	O	O	O
Cross-Docking	O	O	O	O	O	O	O	O	O	O
Warehousing	O	O	O	O	O	O	O	O	O	O
Inventory Control / Management	O	O	O	O	O	O	O	O	O	O
Pick/Pack Operations	O	O	O	O	O	O	O	O	O	O
Assembly	O	O	O	O	O	O	O	O	O	O
Product Returns	O	O	O	O	O	O	O	O	O	O
Logistics Information Systems	O	O	O	O	O	O	O	O	O	O
Lead Logistics Management	O	O	O	O	O	O	O	O	O	O
Consulting services	O	O	O	O	O	O	O	O	O	O

B. Relationship with your logistics service provider

Please indicate your level of agreement with the following statements on how satisfied you are with the relationship between this logistics service provider and your company.	*Strongly Disagree*			*Neutral*			*Strongly Agree*
Please indicate your level of agreement with each statement.							
Our LSP completely fulfills the goals and expectations we jointly set prior to this logistics outsourcing relationship.	O	O	O	O	O	O	O
We are very satisfied with our LSP.	O	O	O	O	O	O	O
The cooperation with our LSP is very good.	O	O	O	O	O	O	O
Our LSP delivers its services always with the required quality.	O	O	O	O	O	O	O
Our LSP delivers its services always in the required time.	O	O	O	O	O	O	O
Through this cooperation, our logistics costs have been reduced to the level we expected.	O	O	O	O	O	O	O
We are very satisfied with the way we interact with our LSP.	O	O	O	O	O	O	O
Differences when cooperating with this LSP are always settled smoothly.	O	O	O	O	O	O	O
The relationship with this LSP is very good.	O	O	O	O	O	O	O
The LSP's own savings from process improvements are passed on to us to a fair extent.	O	O	O	O	O	O	O
This LSP only looks out for itself.	O	O	O	O	O	O	O
We feel fairly treated by this LSP.	O	O	O	O	O	O	O
We and the LSP benefit from this outsourcing relationship to the same extent.	O	O	O	O	O	O	O

Please indicate your level of agreement with the following statements regarding your logistics service provider.	*Strongly Disagree*			*Neutral*			*Strongly Agree*
Please indicate your level of agreement with each statement.							
This LSP keeps promises it makes to our firm.	O	O	O	O	O	O	O
Regarding problems, this LSP is always honest with us.	O	O	O	O	O	O	O
This LSP is trustworthy.	O	O	O	O	O	O	O
This LSP its genuinely concerned that our business succeeds.	O	O	O	O	O	O	O
The LSP correctly carries out tasks that we cannot directly control.	O	O	O	O	O	O	O
We come to our LSP's defense when it is criticized by persons from inside or outside our organization.	O	O	O	O	O	O	O
We would be very sorry personally if we had to terminate the relationship with this LSP.	O	O	O	O	O	O	O
We feel personally offended, when this LSP is criticized by persons from inside or outside of our company.	O	O	O	O	O	O	O
We strongly intend to keep up the relationship with this LSP for as long as possible.	O	O	O	O	O	O	O

Please indicate your level of agreement with the following statements regarding your logistics service provider.	Strongly Disagree			Neutral			Strongly Agree
Please indicate your level of agreement with each statement.							
Our approach to doing business or organizing activities is very similar to our LSP.	O	O	O	O	O	O	O
In the relationship with our LSP, we always pull together in the same direction.	O	O	O	O	O	O	O
When problems or questions arise during this outsourcing relationship, we make decisions together with our LSP to get to adequate solutions.	O	O	O	O	O	O	O
In our business relationship, both parties fully respect each other.	O	O	O	O	O	O	O
Our employees are working together with the LSP to secure the relationship's success even beyond the previously established responsibilities.	O	O	O	O	O	O	O
The LSP is cooperating with us very well.	O	O	O	O	O	O	O
In our relationship with this LSP we frequently run into conflicts on the organizational level.	O	O	O	O	O	O	O
On the operational level, conflicts between our employees and those of the LSP frequently occur.	O	O	O	O	O	O	O
On the whole, our relationship with this LSP is characterized by frequent conflicts.	O	O	O	O	O	O	O
We frequently discuss possible problems or improvements with the responsible parties of our LSP.	O	O	O	O	O	O	O
The exchange of information between the employees of the LSP and our company is working very well.	O	O	O	O	O	O	O
The exchange of information between us and our LSP takes place as soon as it becomes available.	O	O	O	O	O	O	O
Both sides can always fully rely on the information we exchange.	O	O	O	O	O	O	O
The way we exchange information with our LSP is very suited for solving problems according to both parties' interests.	O	O	O	O	O	O	O

Please indicate your level of agreement with the following statements towards your internal relationship and the involvement of your LSP in this outsourcing project. Please indicate your level of agreement with each statement.	*Strongly Disagree*	*Neutral*	*Strongly Agree*
The LSP was significantly involved in the outsourcing relationship at an early stage.	O O O O O O O		
The relationship with our LSP is informal.	O O O O O O O		
When problems or questions arise, we can quickly contact the LSP and jointly find solutions.	O O O O O O O		
The relationship with our LSP is very open.	O O O O O O O		
If one of the parties involved is not fully pleased with something, we immediately and openly talk about it.	O O O O O O O		
The LSP is always open and honest with us.	O O O O O O O		
When problems occur we always solve them jointly.	O O O O O O O		
Especially when solving problems, the exchange of information between the LSP and us is working very well.	O O O O O O O		
Differences of opinion between the LSP and us are viewed as "just part of doing business" and offer potential benefits for both parties involved.	O O O O O O O		

Please indicate your level of agreement with the following statements on the personal relationships of your employees with those of the LSP. Please indicate your level of agreement with each statement.	*Strongly Disagree*	*Neutral*	*Strongly Agree*
There are many personal ties between our employees and those of the LSP.	O O O O O O O		
The relationship with our LSP is working very well on a personal level.	O O O O O O O		
If we were to drop this partner, my colleagues and I would lose good business friends.	O O O O O O O		
We have excellent social relations with this LSP.	O O O O O O O		

Please indicate your level of agreement with the following statements regarding your contact person at the logistics service provider. Please indicate your level of agreement with each statement.	*Strongly Disagree*	*Neutral*	*Strongly Agree*
This person is trustworthy.	O O O O O O O		
This person does not make false promises.	O O O O O O O		
This person is a highly knowledgeable manager.	O O O O O O O		
This person speaks the same "language" we do.	O O O O O O O		
We trust this person to keep our best interests in mind.	O O O O O O O		

What position does your main contact person with the logistics service provider have?	
Please select one of the options or enter the position in the space provided.	
O	President/ CEO/ COO
O	Branch Manager/ Director
O	Division Manager/ Vice President
O	Sales Manager
O	Key Account Manager
O	Others ____________________

Please indicate your level of agreement with the following statements on the continuous improvement efforts of this LSP.	*Strongly Disagree*	*Neutral*	*Strongly Agree*
Please indicate your level of agreement with each statement.			
The LSP puts strong effort into continously optimizing logistics processes.	O O O O O O O		
The LSP continuously makes suggestions for improvements of activities, even those outside its direct responsibility.	O O O O O O O		
When the situation changes, the LSP by itself modifies the logistics activities and processes, if this is useful and necessary.	O O O O O O O		
The LSP shows initiative by approaching us with suggestions for improvement.	O O O O O O O		
The LSP shows a high level of innovation.	O O O O O O O		

Please indicate your level of agreement with the following statements on your satisfaction with the LSP's service quality.	*Strongly Disagree*	*Neutral*	*Strongly Agree*
Please indicate your level of agreement with each statement.			
Overall, this LSP offers excellent services.	O O O O O O O		
This LSP offers great performance.	O O O O O O O		
This LSP offers very high-quality services.	O O O O O O O		
The goals and expectations we jointly set prior to entering this relationship have been significantly exceeded.	O O O O O O O		
We are significantly more satisfied with the quality of the LSP services than we expected.	O O O O O O O		
Our expectations concerning the reduction of costs through this relationship were significantly exceeded.	O O O O O O O		
The relationship between actual costs for provided services and the overall service performance is much better than expected.	O O O O O O O		

Please indicate your level of agreement with the following statements regarding your past and future relationship with this LSP.	*Strongly Disagree*			*Neutral*			*Strongly Agree*
Please indicate your level of agreement with each statement.							
We will continue using this LSP in the future.	O	O	O	O	O	O	O
Right now, we intend to extend existing contracts with this LSP.	O	O	O	O	O	O	O
If we knew then what we know now, we would again select this LSP.	O	O	O	O	O	O	O
When the contract with this LSP ends, we will most likely negotiate a new contract with it, without starting a new bidding process.	O	O	O	O	O	O	O
Within our organization, we have recommended preferential consideration of this LSP for further projects.	O	O	O	O	O	O	O
I often mention this LSP to my co-workers in a positive way.	O	O	O	O	O	O	O
I often recommend this LSP to persons outside my company.	O	O	O	O	O	O	O
We often recommend this LSP.	O	O	O	O	O	O	O

Please indicate your level of agreement with the following statements on your dependence on the LSP.	*Strongly Disagree*			*Neutral*			*Strongly Agree*
Please indicate your level of agreement with each statement.							
We have invested a lot of time that would be wasted if we switched to another LSP.	O	O	O	O	O	O	O
We have made significant investments in resources that would be wasted if we switched to another LSP.	O	O	O	O	O	O	O
We have made significant investments into IT systems that would be wasted if we switched to another LSP.	O	O	O	O	O	O	O
We have significantly invested in the development of our LSP.	O	O	O	O	O	O	O
We have invested a lot of time into the integration of our processes with those of our LSP that would be lost if we switched to another LSP.	O	O	O	O	O	O	O

Please indicate the level of agreement you believe exists between you and your LSP on the following issues:	*Strongly Disagree*	*Neutral*	*Strongly Agree*
If an employee is discovered to have engaged in unethical behavior that results primarily in personal gain, he should be promptly reprimanded.	O O O O O O O		
In business relationships, companies should not only consider their own short-term advantages, but also keep an eye on the long-term benefits for both parties involved.	O O O O O O O		
The employees on both sides have fully understood the goals of the relationship and act accordingly.	O O O O O O O		
The goals of the relationship are defined clearly and are being pursued equally by both parties.	O O O O O O O		
The basic understanding of our relationship is the same for both sides.	O O O O O O O		

Please indicate your level of agreement with the following statements regarding your LSP's pricing.	*Strongly Disagree*	*Neutral*	*Strongly Agree*
Please indicate your level of agreement with each statement.			
Compared to other LSPs, our LSP's prices are very good.	O O O O O O O		
The LSP offers a very good price-performance ratio.	O O O O O O O		
Compared to carrying out those tasks ourselves, our LSP's prices are very good.	O O O O O O O		

Please indicate your level of agreement with the following statements regarding your past and future relationship with this LSP.	*Strongly Disagree*	*Neutral*	*Strongly Agree*
Please indicate your level of agreement with each statement.			
In the future, the LSP will have a higher share of our logistics volume.	O O O O O O O		
In the future, we will use this LSP more than we do now.	O O O O O O O		
When we bid out other services than the ones we outsource today, we will consider this LSP preferentially.	O O O O O O O		
When we outsource additional services, we will first offer them to this LSP, before starting a bidding process.	O O O O O O O		

Please indicate your level of agreement with the following statements on your relationship with this LSP.	*Strongly Disagree*	*Neutral*	*Strongly Agree*
Please indicate your level of agreement with each statement.			
To accomplish its own objectives, sometimes our LSP alters the facts slightly.	O O O O O O O		
To accomplish its own objectives, sometimes our LSP promises to do things without actually doing them later.	O O O O O O O		
Our LSP sometimes exaggerates its requirements in order to get what it really needs from us.	O O O O O O O		
Our LSP feels that it is alright to do anything within its means to further its own interests.	O O O O O O O		
Our LSP feels that honesty does pay when dealing with us.	O O O O O O O		

Please indicate your level of agreement with the following statements on your attitude towards this LSP and on market alternatives.	*Strongly Disagree*	*Neutral*	*Strongly Agree*
Please indicate your level of agreement with each statement.			
We cannot terminate the relationship to this LSP, because we have invested too much.	O O O O O O O		
It would be difficult to terminate the relationship with this LSP, even if we wanted to.	O O O O O O O		
We feel obligated to keep up the relationship with this LSP.	O O O O O O O		
We cannot consider terminating the relationship to this LSP, because the LSP has invested too much into the relationship.	O O O O O O O		
There are many alternative LSPs that can offer the same service at the same cost.	O O O O O O O		

Please indicate your level of agreement with the following statements on the specific investments your LSP had to make for this relationship.	*Strongly Disagree*	*Neutral*	*Strongly Agree*
Please indicate your level of agreement with each statement.			
To supply us with logistics services, our LSP had to make significant investments (transportation equipment, warehouses, etc.) that it can only use for us.	O O O O O O O		
The investments our LSP had to make in employees, processes or assets will quickly amortize.	O O O O O O O		

Attention: Please answer the following questions with regard to your company and your customers.

C. Questions on the present situation of your company

Please indicate your level of agreement with the following statements about the present situation of your company.	*Strongly Disagree*			*Neutral*			*Strongly Agree*
Please indicate your level of agreement with each statement.							
We find it very difficult to predict the future needs and requirements of our customers.	O	O	O	O	O	O	O
Our customers vary significantly in size, resulting in considerably different sales volumes.	O	O	O	O	O	O	O
The orders of our customers are very unevenly distributed and differ depending on the specific order or season.	O	O	O	O	O	O	O
The pressure of competition in our industry is very high.	O	O	O	O	O	O	O
Our customers increasingly view a very good logistics performance as important, making it a competitive advantage for us.	O	O	O	O	O	O	O
Our product range is very broad with a lot of different products.	O	O	O	O	O	O	O
Compared to our competitors, our products are relatively higher price.	O	O	O	O	O	O	O

Please indicate your level of agreement with the following statements about your internal logistics organization.	*Strongly Disagree*			*Neutral*			*Strongly Agree*
Please indicate your level of agreement with each statement.							
The decision-making process on logistics issues is very centralized in our company.	O	O	O	O	O	O	O
Our Business Unit has a very smooth, continuous, quick and largely failure-free flow of material and information.	O	O	O	O	O	O	O
Our Business Unit is managed in a very flow- or process-oriented way.	O	O	O	O	O	O	O
Our Business Unit's degree of flow or process orientation is significantly higher than that of our competitors.	O	O	O	O	O	O	O
All business processes are very well coordinated.	O	O	O	O	O	O	O
In our Business Unit a large number of individual interests exists that impede the direct achievement of all our objectives.	O	O	O	O	O	O	O

D. Questions on your logistics success and overall success

How does your company's performance on logistics cost compare to your competitors?	*A lot worse*	*About the same*	*A lot better*
Please rate each factor.			
Logistics costs relative to total revenue (compensation for the LSP included).	O O O O O O O		
Warehousing costs.	O O O O O O O		
Transportation costs.	O O O O O O O		
IT costs in support of logistics activities.	O O O O O O O		
Human resource costs of logistics activities.	O O O O O O O		

How does your company's performance on logistics service compare to your competitors?	*A lot worse*	*About the same*	*A lot better*
Please rate each factor.			
Order lead times.	O O O O O O O		
Delivery times.	O O O O O O O		
Ability for on-time delivery.	O O O O O O O		
Delivery reliability.	O O O O O O O		
Degree of delivery flexibility.	O O O O O O O		
Degree of damage- and error-free logistical activities.	O O O O O O O		

How does your company's market performance compare to your competitors?	*A lot worse*	*About the same*	*A lot better*
Please rate each factor.			
Customer satisfaction.	O O O O O O O		
Value added provided to the customer.	O O O O O O O		
Retention of existing customers.	O O O O O O O		
Acquisition of new customers.	O O O O O O O		
Achieving our desired growth rate.	O O O O O O O		
Achieving our desired market share.	O O O O O O O		
Adaptation of products / services to new customer requirements.	O O O O O O O		
Reacting to new developments in the market.	O O O O O O O		
Benefiting from new opportunities in the market.	O O O O O O O		

How does your company's financial performance compare to your competitors?	*A lot worse*	*About the same*	*A lot better*
Our return on sales last year with respect to our competition was...	O O O O O O O		
Our average return on sales over the past three years with respect to our competition was...	O O O O O O O		
The development of our return on sales during the past three years with respect to our competition was...	O O O O O O O		

Thank you for completing the survey to this point.

We would now like to ask you to complete a few background questions. As with the rest of the survey, we guarantee the strictest confidentiality!

What is your position?	
O	President/ CEO/ COO
O	Vice President Logistics
O	Logistics Manager
O	Logistics Employee
O	Others ____________________

For how many years have you been working in this position?	for ____ years (e.g. 2.5)

For how many years have you been working for this company?	for ____ years (e.g. 2.5)

Which industry does your business unit belong to?	
Please select only one industry.	
O	Food and beverage
O	Automotive
O	Consumer goods
O	Industrial equipment
O	Electronics and related instruments
O	Computer hardware and peripheral equipment
O	Chemicals and plastics
O	Retailing
O	Healthcare
O	Others ____________________

What are the current yearly revenues of your business unit (in $ Million/year)?							
up tp 24	O	25-49	O	50-99	O	100-249	O
250-499	O	500-999	O	1,000-4,999	O	More than 5,000	O

What is the approximate number of employees in your business unit?	________ employees
How many LSPs does your business unit use?	________ logistics service providers
What is the share of logistics cost that external LSPs have (in %)?	________ percent

Please complete the following questions having your selected LSP in mind.

What is the share of total outsourced logistics cost that this LSP has?	________ percent
For how long have you been working together with this LSP?	for _____ years (e.g. 2.5)
For how long have you been working together with this LSP in a way that you would call a "close relationship"?	for _____ years (e.g. 2.5)
What is the total duration of the current contract with this LSP?	_____ years (e.g. 2.5)
What is the remaining duration of this contract with this LSP?	_____ years (e.g. 2.5)

Thank you for taking part in the survey!

Please provide us with the address to which we may forward your complimentary report and book and notify you should you win the iPod raffle.

Last name: ______________________________

First name: ______________________________

Email address: ______________________________

Company: ______________________________

Street: ______________________________

Zip: ______________________________

City: ______________________________

Phone number: ______________________________

Please use the space provided for comments and suggestions:

List of figures

List of tables

References

Aaker, D. A. (1996): Measuring brand equity across products and markets, in: California Management Review 38(3): 102-120.

Adams, J. S. (1963): Toward an understanding of inequity, in: Journal of Abnormal and Social Psychology 5: 422-436.

Adams, J. S. (1965): Inequity in Social Exchange, in: Advances in Experimental Social Psychology 2: 267-299.

Adler, N. J., Doktor, R., and Redding, S. G. (1986): From the Atlantic to the Pacific Century: Cross-Cultural Management Reviewed, in: Journal of Management 12(2): 295-318.

Adler, N. J. and Graham, J. L. (1989): Cross-Cultural Interaction: The International Comparison Fallacy?, in: Journal of International Business Studies Fall: 515-537.

Africk, J. M. and Calkins, C. S. (1994): Does asset ownership mean better service?, in: Transportation & Distribution 35(5): 49-61.

Agustin, C. and Singh, J. (2005): Curvilinear Effects of Consumer Loyalty Determinants in Relational Exchanges, in: Journal of Marketing Research 42(1): 96-108.

Algesheimer, R., Dholakia, U. M., and Herrmann, A. (2005): The Social Influence of Brand Community: Evidence from European Car Clubs, in: Journal of Marketing 69(3): 19-34.

Allen, J. A. and Meyer, J. P. (1990): The Measurement and Antecedents of Affective, Continuance and Normative Commitment to the Organization, in: Journal of Occupational Psychology 63(1): 75-99.

Andersen, P. H. (2001): Relationship development and marketing communication: an integrative model, in: Journal of Business & Industrial Marketing 3: 167-182.

Anderson, D. R., Burnham, K. P., and White, G. C. (1998): Comparison of Akaike information criterion and consistent Akaike information criterion for model selection and statistical inference from capture-recapture studies, in: Journal of Applied Statistics 25(2): 263-282.

Anderson, E. and Weitz, B. (1992): The use of pledges to build and sustain commitment in distribution channels, in: Journal of Marketing Research 29(1): 18-34.

Anderson, E. W., Fornell, C., and Lehmann, D. R. (1994): Customer Satisfaction, Market Share, and Profitability: Findings from Sweden, in: Journal of Marketing 58: 53-66.

Anderson, E. W. and Sullivan, M. W. (1993): The Antecedents and Consequences of Customer Satisfaction for Firms, in: Marketing Science 12(2): 125-143.

Anderson, J. C. and Gerbing, D. W. (1988): Structural Equation Modelin in Practice: A Review and Recommended Two-Step Approach, in: Psychological Bulletin(3): 411-423.

Anderson, J. C., Gerbing, D. W., and Hunter, J. E. (1987): On the Assessment of Unidimensional Measurement: Internal and External Consistency, and Overall Consistency Criteria, in: Journal of Marketing Research 24: 432-437.

Anderson, J. C., Hakansson, H., and Johanson, J. (1994): Dyadic business relationships within a business network context, in: Journal of Marketing 58(4): 1-15.

Anderson, J. C. and Narus, J. A. (1984): A Model of the Distributor's Perspective of Distributor-Manufacturer Working Relationships, in: Journal of Marketing 48: 62-74.

Anderson, J. C. and Narus, J. A. (1990): A Model of Distributor Firm and Manufacturer Firm Working Partnerships, in: Journal of Marketing 54(1): 42-58.

Aristotle (1999): Nicomachean ethics, Indianapolis.

Armstrong, J. S. and Overton, T. S. (1977): Estimating Nonresponse Bias in Mail Surveys, in: Journal of Marketing Research 14: 396-402.

Arrow, K. J. (1975): Gifts and Exchanges, in: Phelps, E. S., Altruism, Morality, and Economic Theory, New York: 13-28.

Austen, S. (2002): An International Comparison of Attitudes to Inequality, in: International Journal of Social Economics 29(3-4): 218-237.

Austin, W. and Walster, E. (1975): Equity With The World: The Trans-Relational Effects of Equity and Inequity, in: Socionometry 38(4): 474-496.

Babakus, E. and Boller, G. W. (1992): An Empirical Assessment of the SERVQUAL Scale, in: Journal of Business Research 24: 253-268.

Babbar, S. and Prasad, S. (1998): International purchasing, inventory management and logistics research, in: International Journal of Physical Distribution & Logistics Management 28(6): 403-433.

Backhaus, K., Baumeister, C., and Mühlfeld, K. (2003): Kundenbindung im Industriegütermarketing, in: Bruhn, M. and Homburg, C., Handbuch Kundenbindungsmanagement: Strategien und Instrumente für ein erfolgreiches CRM, Wiesbaden: 193-221.

Bagozzi, R. P. (1980): Causal Models in Marketing, New York et al.

Bagozzi, R. P. (1981): An Examination of the Validity of two Models of Attitude, in: Mulitvariate Behavioral Research 16: 323-359.

Bagozzi, R. P. (1998): A Prospectus for Theory Construction in Marketing: Revisited and Revised, in: Hildebrandt, L., Homburg, C., Die Kausalanalyse: Ein Instrument der empirischen betriebswirtschaftlichen Forschung, Stuttgart: 45-81.

Bagozzi, R. P. and Baumgartner, H. (1994): The Evaluation of Structural Equation Models and Hypothesis Testing, in: Bagozzi, R. P., Principles of Marketing Research, Cambridge: 386-422.

Bagozzi, R. P. and Fornell, C. (1982): Theoretical Concepts, Measurement, and Meaning, in: Fornell, C., A Second Generation of Multivariate Analysis, New York: 24-38.

Bagozzi, R. P. and Phillips, L. W. (1982): Representing and Testing Organizational Theories: A Holistic Construal, in: Administrative Science Quarterly 27: 459-489.

Bagozzi, R. P. and Yi, Y. (1988): On the Evaluation of Structural Equation Models, in: Journal of the Academy of Marketing Science 16: 74-94.

Bagozzi, R. P., Yi, Y., and Phillips, L. W. (1991): Assessing Construct Validity in Organizational Research, in: Administrative Science Quarterly 36: 421-458.

Baldinger, A. L. and Rubinson, J. (1997): The Jeopardy in Double Jeopardy, in: Journal of Advertising Research 37(6): 37-49.

Ballou, R. H. (2003): Business Logistics Management, 5. Edition, Englewood Cliffs.

Bansal, H. S., Irving, P. G., and Taylor, S. F. (2004): A Three-Component Model of Customer Commitment to Service Providers, in: Journal of the Academy of Marketing Science 32(3): 234-250.

Bardi, E. J. and Tracey, M. (1991): Transportation Outsourcing: A Survey of US Practices, in: International Journal of Physical Distribution & Logistics Management 21(3): 15-21.

Barks, J. V. (1994): Logistics for Hire, in: International Business 7(5): 36-38.

Baruch, Y. (1999): Response Rate in Academic Studies: A Comparative Analysis, in: Human Relations 52(4): 421-438.

Bauer (2000): Kundenzufriedenheit in industriellen Geschäftsbeziehungen, Wiesbaden.

Beerli, A., Martín, J. D., and Quintana, A. (2004): A model of customer loyalty in the retail banking market, in: European Journal of Marketing 38(1/2): 253-275.

Bell, S. J., Auh, S., and Smalley, K. (2005): Customer Relationship Dynamics: Service Quality and Customer Loyalty in the Context of Varying Levels of Customer Expertise and Switching Costs, in: Journal of the Academy of Marketing Science 33(2): 169-183.

Bendapudi, N. and Berry, L. L. (1997): Customer's Motivations for Maintaining Relationships with Service Providers, in: Journal of Retailing 73(1): 15.

Bentler, P. M. and Bonett, D. (1980): Significance Tests and Goodness of Fit in the Analysis of Covariance Structures, in: Psychological Bulletin 88(3): 588-606.

Bentler, P. M. and Chou, C.-P. (1987): Practical Issues in Structural Equation Modeling, in: Sociological Methods & Research 16: 78-117.

Berglund, M., van Laarhoven, P., Sharman, G., and Wandel, S. (1999): Third-Party Logistics: Is There a Future?, in: International Journal of Logistics Management 10(1): 59-70.

Bhatnagar, R., Sohal, A. S., and Millen, R. (1999): Third party logistics services: a Singapore perspective, in: International Journal of Physical Distribution & Logistics Management 29(9/10): 569-587.

Bienstock, C. C., Mentzer, J. T., and Bird, M. M. (1997): Measuring Physical Distribution Service Quality, in: Journal of the Academy of Marketing Science 25(1): 31-44.

Bing, J. W. (2004): Hofstede's consequences: The impact of his work on consulting and business practices, in: Academy of Management Executive 18(1): 80-87.

Biong, H. (1993): Satisfaction and Loyalty to Suppliers within the Grocery Trade, in: European Journal of Marketing 27(7): 21-38.

Bitner, M. J. (1990): Evaluating Service Encounters: The Effects of Physical Surroundings and Employee Responses, in: Journal of Marketing 54(2): 69-82.

Blalock, H. M. J. (1963): Making Causal Inferences for Unmeasured Variables from Correlations among Indicators, in: The American Journal of Sociology 69: 53-62.

Blau, P. M. (1960): A Theory of Social Integration, in: The American Journal of Sociology 65(6): 545-556.

Blau, P. M. (1964): Exchange and Power in Social Life, New York.

Blau, P. M. (1968): Interaction: Social Exchange, in: International Encyclopedia of the Social Sciences 7(2000): 452-458.

Bloemer, J. and de Ruyter, K. (1999): Customer Loyalty in High and Low Involvement Service Settings: The Moderating Impact of Positive Emotions, in: Journal of Marketing Management 15(4): 315-330.

Bloemer, J., de Ruyter, K., and Wetzels, M. (1999): Linking perceived service quality and service loyalty: A multi-dimensional perspective, in: European Journal of Marketing 33(11/12): 1082-1106.

Bloemer, J. and Kasper, H. D. P. (1995): The complex relationship between consumer satisfaction and brand loyalty, in: Journal of Economic Psychology 16(2): 311-329.

Bloemer, J. and Lemmink, J. (1992): The Importance of Customer Satisfaction in Explaining Brand and Dealer Loyalty, in: Journal of Marketing Management 8(4): 351-363.

Bollen, K. A. (1989): Structural Equations with Latent Variables, New York.

Bolton, R. N. (1998): A Dynamic Model of the Duration of the Customer's Relationship With a Continuous Service Provider: The Role of Satisfaction, in: Marketing Science 17(1): 45-65.

Bolton, R. N., Kannan, R. K., and Bramlett, M. D. (2000): Implications of Loyalty Program Membership and Service Experiences for Customer Retention and Value, in: Journal of the Academy of Marketing Science 28(1): 95-108.

Bolumole, Y. A. (2001): The Supply Chain Role of Third-Party Logistics Providers, in: International Journal of Logistics Management 12(2): 87-102.

Bookbinder, J. H. and Tan, C. S. (2003): Comparison of Asian and European logistics systems, in: International Journal of Physical Distribution & Logistics 33(1): 36-58.

Boomsma, A. (1985): Nonconvergence, improper solutions, and starting values in LISREL maximum likelihood estimation, in: Psychometrika 52: 229-242.

Borle, S., Boatwright, P., Kadane, J. B., Nunes, J. C., and Shmueli, G. (2005): The Effect of Product Assortment Changes on Customer Retention, in: Marketing Science 24(4): 616-622.

Boulding, W., Kalra, A., Staelin, R., and Zeithaml, V. A. (1993): A dynamic process model of service quality: From expectations to behavioral intentions, in: Journal of Marketing Research 30(1): 7-27.

Bourdieu, P. (1980): Le sens pratique, Paris.

Bovet, D. (1991): Logistics strategies for Europe in the nineties, in: Transportation & Distribution 19(4): 12-15; 46-48.

Bowersox, D. J., Closs, D. J., and Helferich, O. K. (1986): Logistical Management, A Systems Integration of Physical Distribution, Manufacturing Support and Materials Procurement, New York, London.

Bowersox, D. J. and Daugherty, P. J. (1987): Emerging Patterns of Logistical Organization, in: Journal of Business Logistics 8(1): 46-60.

Boyson, S., Corsi, T. M., Dresner, M., and Rabinovich, E. (1999): Managing Effective Third Party Logistics Relationships: What does it take?, in: Journal of Business Logistics 20(1): 73-98.

Bozdogan, H. (1987): Model Selection and Akaike's Information Criterion (AIC): The General Theory and its Analytical Extension, in: Psychometrika 52(3): 345-370.

Bradley, P. (1994a): Contract logistics: it's all about costs, in: Purchasing 117(6): 56A3-A14.

Bradley, P. (1994b): What really matters, in: Purchasing 117(1): 66-71.

Brady, M. K. and Cronin, J. J. (2001): Some New Thoughts on Conceputalizing Perceived Service Quality: A Hierarchical Approach, in: Journal of Marketing 65: 34-49.

Brady, M. K., Cronin, J. J., and Brand, R. R. (2002): Performance-only measurement of service quality: a replication and extension, in: Journal of Business Research 55: 17-31.

Braunstein, C. (2001): Einstellungsforschung und Kundenbindung, Wiesbaden.

Bretzke, W.-R. (1993): Pro und contra Outsourcing von Logistikdienstleistungen, in: Beschaffung aktuell 1993(6): 37-39.

Brown, T. J., Churchill Jr., G. A., and Peter, J. P. (1993): Research Note: Improving the Measurement of Service Quality, in: Journal of Retailing 69(1): 127-139.

Browne, M. and Allen, J. (2001): Logistics out-sourcing, in: Brewer, A. M., Handbook of Logistics and Supply Chain Management, Oxford: 253-268.

Browne, M. W. (2001): An Overview of Analytic Rotation in Exploratory Factor Analysis, in: Multivariate Behavioral Research 36(1): 111-150.

Browne, M. W. and Cudeck, R. (1989): Single Sample Cross-Validation Indices for Covariance Structures, in: Multivariate Behavioral Research 24(4): 445-455.

Browne, M. W. and Cudeck, R. (1993): Alternative Ways of Assessing Model Fit, in: Bollen, K. A. and Long, J. S., Testing Structural Equation Models, Newbury Park: 136-162.

Byrne, P. M. (1993): A new road map for contract logistics, in: Transportation & Distribution 34(4): 58-62.

Byrne, P. M. (2001): Structural Equation Modeling with AMOS: Basic Concepts, Applications and Programming, Mahwah.

Cadotte, E. R., Woodruff, R. B., and Jenkins, R. L. (1987): Expectations and Norms in Models of Consumer Satisfaction, in: Journal of Marketing Research 24: 305-314.

Calfee, J. E. and Rubin, P. H. (1993): Nontransactional Data in Managerial Economics and Marketing, in: Managerial & Decision Economics 14(2): 163-173.

Candler, J. (1994): You make it, they distribute it, in: Nation's Business 82(3): 46-48.

Carter, C. R., Ellram, L. M., and Ready, K. J. (1998): Environmental Purchasing: Benchmarking Our German Counterparts, in: International Journal of Purchasing & Materials Management 34(4): 28-38.

Caruana, A. (2000): Service Loyalty: The effects of service quality and the mediating role of customer satisfaction, in: European Journal of Marketing 36: 811-828.

Caruana, A., Money, A. H., and Berthon, P. R. (2000): Service quality and satisfaction - the moderating role of value, in: European Journal of Marketing 34: 1338-1352.

Cavinato, J. (1999): A general methodology for determining a fit between supply chain logistics and five stages of strategic management, in: International Journal of Physical Distribution & Logistics 29: 162-180.

Chaudhuri, A. and Holbrook, M. B. (2001): The Chain of Effects from Brand Trust and Brand Affect to Brand Performance: The Role of Brand Loyalty, in: Journal of Marketing 65(2): 81-93.

Child, J. (1972): Organization Structure and Strategies of Control: A Replication of the Aston Study, in: Administrative Science Quarterly 17(2): 163-177.

Chou, C.-P. and Bentler, P. M. (1995): Estimation and Tests in Structural Equation Modeling, in: Hoyle, R. H., Structural Equation Modeling: Concepts, Issues, and Applications, London, New Delhi: 37-55.

Chow, G., Heaver, T. D., and Henriksson, L. E. (1994): Logistics Performance: Definition and Measurement, in: International Journal of Physical Distribution & Logistics Management 24(1): 17-28.

Churchill, G. A. (1979): A Paradigm for Developing Better Measures of Marketing Constructs, in: Journal of Marketing Research 16: 64-73.

Churchill, G. A. and Suprenant, C. (1982): An Investigation into the Determinants of Customer Satisfaction, in: Journal of Marketing Research 19: 491-504.

Coase, R. H. (1937): The Nature of the Firm, in: Economica 4: 386-405.

Cook, K. S. and Emerson, R. M. (1978): Power, Equity and Commitment in Exchange Networks, in: American Sociological Review 43(5): 721-739.

Cooke, J. A. (2000): 3PLs: Riding the wave, in: Logistics Management & Distribution Report 39(7): 69-71.

Cooper, J. C. (1993): Logistics strategies for global businesses, in: International Journal of Physical Distribution & Logistics Management 23(4): 12-23.

Copeland, M. T. (1923): Relation of Consumers' Buying Habits to Marketing Methods, in: Harvard Business Review 1(3): 282-289.

Cronbach, L. J. (1951): Coefficient Alpha and the Internal Structure of Tests, in: Psychometrika 16: 297-334.

Cronin Jr., J. J. and Taylor, S. A. (1992): Measuring service quality: A reexamination and extension, in: Journal of Marketing 56(3): 55-68.

Crutchfield, T. N. (2001): The Effect of Trust and Commitment on Retention of High-Risk Professional Service Customers, in: Services Marketing Quarterly 22(2): 17-27.

CSCMP (2005): Council of Supply Chain Management Professionals, http://cscmp.org/Website/AboutCSCMP/Definitions/Definitions.asp, Sept. 14, 2005.

Cunningham, R. M. (1956): Brand Loyalty - what, where, how much?, in: Harvard Business Review 34(1): 116-128.

Dabholkar, P. A. (1995): The Convergence of Customer Satisfaction and Service Quality Evaluations with increasing Customer Patronage, in: Journal of Satisfaction 8(2): 32-43.

Dabholkar, P. A., Shepherd, C. D., and Thorpe, D. I. (2000): A Comprehensive Framework for Service Quality: An Investigation of Critical Conceptual and Measurement Issues Through a Longitudinal Study, in: Journal of Retailing 76: 139-173.

D'Andrade, R. G. (1984): Cultural Meaning Systems, in: Shweder, R. A. and LeVine, R. A., Culture Theory: Essay on Mind, Self, and Emotion, Cambridge, MA: 88-119.

Day, G. S. (1969): A Two-Dimensional Concept Of Brand Loyalty, in: Journal of Advertising Research 9(3): 29-35.

de Ruyter, K. and Bloemer, J. (1999): Customer loyalty in extended service settings, in: International Journal of Service Industry Management 10(3): 320-336.

de Ruyter, K., Wetzels, M., and Bloemer, J. (1998): On the relationship between perceived service quality, service loyalty and switching costs, in: International Journal of Service Industry Management 9(5): 436-453.

De Wulf, K., Odekerken-Schröder, G., and Iacobucci, D. (2001): Investments in Consumer Relationships: A Cross-Country and Cross-Industry Exploration, in: Journal of Marketing 65(4): 33-40.

Deepen, J. (2003): Die Rolle von Logistikdienstleistern im Supply Chain Management - Eine konzeptionell-empirische Betrachtung, in: Weber, J. and Deepen, J., Erfolg durch Logistik: Erkenntnisse aktueller Forschung, Bern, Stuttgart, Vienna: 115-165.

Deepen, J. (2006): Logistics Outsourcing Relationships: Measurement, Antecedents, and Effects of Logistics Outsourcing Performance, Forthcoming.

Dehler, M. (2001): Entwicklungsstand der Logistik: Messung - Determinanten - Erfolgswirkungen: Erkenntnisse aktueller Forschung, Wiesbaden.

Deshpandé, R., Farley, J. U., and Webster Jr., F. E. (1993): Corporate Culture Customer Orientation, and Innovativeness in Japanese Firms: A Quadrad Analysis, in: Journal of Marketing 57(1): 23-37.

Diamantopoulos, A. and Winklhofer, H. M. (2001): Index Construction with Formative Indicators: An Alternative to Scale Development, in: Journal of Marketing Research (JMR) 38(2): 269-277.

Dick, A. S. and Basu, K. (1994): Customer Loyalty: Toward an Integrated Conceptual Framework, in: Journal of the Academy of Marketing Science 22(2): 99-113.

Diller, H. (1995): Kundenbindung als Zielvorgabe im Beziehungs-Marketing, Arbeitspapier Nr. 40, Universität Erlangen-Nürnberg, Lehrstuhl für Marketing, Nürnberg.

Doney, P. M. and Cannon, J. P. (1997): An Examination of the Nature of Trust in Buyer-Seller Relationships, in: Journal of Marketing 61: 35-51.

Doney, P. M., Cannon, J. P., and Mullen, M. R. (1998): Understanding the Influence of National Culture on the Development of Trust, in: Academy of Management Review 23(3): 601-620.

Dunn, S. C., Seaker, R. F., and Waller, M. A. (1994): Latent Variables in Business Logistics Research: Scale Development and Validation, in: Journal of Business Logistics 15(2): 145-172.

Durvasula, S., Andrews, J. C., Lysonski, S., and Netemeyer, R. G. (1993): Assessing the Cross-national Applicability of Consumer Behavior Models: A Model of Attitude toward Advertising in General, in: Journal of Consumer Research 19(4): 626-636.

Dwyer, F. R., Schurr, P. H., and Oh, S. (1987): Developing Buyer-Seller Relationships, in: Journal of Marketing 51(2): 11-27.

Eckhardt, G. (2002): Culture's Consequences: Comparing Values, Behaviors, Institutions and Organisations Across Nations, in: Australian Journal of Management 27(1): 89-94.

Emerson, R. M. (1962): Power-Dependence Relations, in: American Sociological Review 27(2): 31-41.

Engelbrecht, C. (2003): Logistik-Outsourcing: Erfolgsfaktoren und Erfolgswirkung - Erkenntnisse aus der Praxis, in: Weber, J. and Deepen, J., Erfolg durch Logistik: Erkenntnisse aktueller Forschung, Bern, Stuttgart, Vienna: 43-76.

Engelbrecht, C. (2004): Logistikoptimierung durch Outsourcing: Erfolgswirkung und Erfolgsfaktoren, Wiesbaden.

Ernst, H. (2001): Erfolgsfaktoren neuer Produkte - Grundlagen für eine valide empirische Forschung, Wiesbaden.

Ernst, H. (2003): Ursachen eines Informant Bias und dessen Auswirkung auf die Validitat empirischer betriebswirtschaftlicher Forschung, in: Zeitschrift für Betriebswirtschaft 73(12): 1249-1275.

Fantasia, J. J. (1993): Are you a candidate for third party logistics?, in: Transportation & Distribution 34(1): 30.

Fassnacht, M. and Daus, P. W. (2004): Loyalty Determinants - Literature Review and Development of an Integrative Model of Customer Loyalty, Research Paper Nr. 26, Center for Market-Orientated Management (ZMU), WHU - Otto Beisheim School of Management, Vallendar.

Fawcett, S. E., Birou, L., and Taylor, B. C. (1993): Supporting global operations through logistics and purchasing, in: International Journal of Physical Distribution & Logistics Management 23(4): 3-11.

Fawcett, S. E. and Clinton, S. R. (1997): Enhancing Logistics to Improve the Competitiveness of Manufacturing Organizations: A Triad Perspective, in: Transportation Journal 37(1): 18-28.

Ford, J. D. and Slocum Jr., J. W. (1977): Size, Technology, Environment and the Structure of Organizations, in: Academy of Management Review 2(4): 561-575.

Fornell, C. (1992): A National Customer Satisfaction Barometer: The Swedish Experience, in: Journal of Marketing 56(1): 6-21.

Fornell, C., Johnson, M. D., Anderson, E. W., Cha, J., and Bryant, B. E. (1996): The American customer satisfaction index: Nature, purpose, and findings, in: Journal of Marketing 60(4): 7-18.

Fornell, C. and Larcker, D. F. (1981): Evaluating Structural Equation Models with Unobservable Variables and Measurement Error, in: Journal of Marketing Research (JMR) 18(1): 39-50.

Foster, T. A. and Muller, E. J. (1990): Third parties: your passport to profits, in: Distribution 89(10): 30-32.

Fournier, S. and Yao, J. L. (1997): Reviving brand loyalty: A reconceptualization within the framework of consumer - brand relationships, in: International Journal of Research in Marketing 14(5): 451-472.

Fullerton, G. (2005): How commitment both enables and undermines marketing relationships, in: European Journal of Marketing 39(11/12): 1372-1388.

Furrer, O., Liu, B. S.-C., and Sudharshan, D. (2000): The Relationships Between Culture and Service Quality Perceptions: Basis for Cross-Cultural Market Segmentation and Resource Allocation, in: Journal of Service Research 2(4): 355-371.

Gallagher, D. (1996): Book review: Fons Trompenaars. Riding the waves of culture: Understanding diversity in global business, in: Personnel Psychology 49(1): 228-231.

Ganesan, S. (1994): Determination of Long-Term Orientation in Buyer-Seller Relationships, in: Journal of Marketing 58(4): 1-19.

Garbarino, E. and Johnson, M. S. (1999): The Different Roles of Satisfaction, Trust and Commitment in Customer Relationships, in: Journal of Marketing 63: 70-87.

Garver, M. S. and Mentzer, J. T. (1999): Logistics Research Methods: Employing Structural Equation Modeling to Test for Construct Validity, in: Journal of Business Logistics 20: 33-57.

Gecker, R. (2004): The State of Logistics Goes Global, in: Inbound Logistics 24(7): 16-20.

Gerbing, D. W. and Anderson, J. C. (1988): An Updated Paradigm for Scale Development Incorporating Unidimensionality and Its Assessment, in: Journal of Marketing Research 25: 186-192.

Giering, A. (2000): Der Zusammenhang zwischen Kundenzufriedenheit und Kundenloyalität - eine Untersuchung moderierender Effekte, Wiesbaden.

Giese, J. L. and Cote, J. A. (2000): Defining Consumer Satisfaction, in: Academy of Marketing and Science Review 1: www.amsreview.org/articles/giese01-2000.pdf.

Goldberg, D. (1990): JIT's next step: moves cargo and data, in: Transportation & Distribution 31(13): 18-22.

Golob, T. F. (2003): Structural Equation Modeling for Travel Behavior Research, in: Transportation Research - Part B: Methodological 37(1): 1-25.

Goodenough, W. H. (1981): Culture, Language, and Society, Mento Park.

Goodstein, L. D. (1981): Commentary: Do American theories apply abroad?, in: Organizational Dynamics 10(1): 49-54.

Gorsuch, R. L. (1997): Exploratory Factor Analysis: Its Role in Item Analysis, in: Journal of Personality Assessment 68(3): 532.

Grant, D. (2004): UK and US management styles in logistics: different strokes for different folks?, in: International Journal of Logistics: Research & Applications 7(3): 181-197.

Grant, D., Teller, C., and Teller, W. (2005): Web-based Surveys in Logistics Research: An Empirical Application, in: Kotzab, H., Seuring, S., Müller, M., and Reiner, G., Research Methodologies in Supply Chain Management, Heidelberg, New York: 139-154.

Granzin, K. L. and Painter, J. J. (2001): Motivational Influences on 'Buy Domestic' Purchasing: Marketing Management Implications from a Study of Two Nations, in: Journal of International Marketing 9(2): 73-96.

Greenberg, J. (2001): Studying Organizational Justice Cross-Culturally: Fundamental Challenges, in: International Journal of Conflict Management 12(4): 365-367.

Griesinger, D. W. (1990): The Human Side of Economic Organization, in: Academy of Management Review 15(3): 478-499.

Griffis, S. E., Goldsby, T. J., and Cooper, M. (2003): Web-Based and Mail Surveys: A Comparison of Response, Data and Cost, in: Journal of Business Logistics 24(2): 237-258.

Gronholdt, L., Martensen, A., and Kristensen, K. (2000): The relationship between customer satisfaction and loyalty: cross-industry differences, in: Total Quality Management 11(4/5/6): S509-S514.

Grönroos, C. (1989): Defining Marketing: A Market-Oriented Approach, in: European Journal of Marketing 23(1): 52-60.

Grönroos, C. (1994): Quo Vadis, Marketing? Toward a Relationship Marketing Paradigm, in: Journal of Marketing Management 10: 347-360.

Grönroos, C. (1995): Relationship Marketing: The Strategy Continuum, in: Journal of the Academy of Marketing Science 23(4): 252-254.

Grönroos, C. (1997): From marketing mix to relationship marketing - towards a paradigm shift in marketing, in: Management Decision 35(3/4): 322-339.

Gruen, T. W., Summers, J. O., and Acito, F. (2000): Relationship Marketing Activities, Commitment, and Membership Behaviors in Professional Associations, in: Journal of Marketing 64(7): 34-49.

Guenzi, P. and Pelloni, O. (2004): The impact of interpersonal relationships on customer satisfaction and loyalty to the service provider, in: International Journal of Service Industry Management 15(4): 365-384.

Gujarati, D. N. (2003): Basic Econometrics, 4. Edition, London.

Gummesson, E. (1997): Relationship marketing as a paradigm shift: some conclusions from the 30R approach, in: Management Decision 35: 267-272.

Gundlach, G. T., Achrol, R. S., and Mentzer, J. T. (1995): The Structure of Commitment in Exchange, in: Journal of Marketing 59(1): 78-92.

Gupta, S., Lehmann, D. R., and Stuart, J. A. (2004): Valuing Customers, in: Journal of Marketing Research (JMR) 41(1): 7-18.

Gustafsson, A., Johnson, M. D., and Roos, I. (2005): The Effects of Customer Satisfaction, Relationship Commitment Dimensions, and Triggers on Customer Retention, in: Journal of Marketing 69(4): 210-218.

Hall, E. T. (1960): The Silent Language in Overseas Business, in: Harvard Business Review 38(3): 87-96.

Hall, E. T. (1976): Beyond Culture, Graden City.

Hall, E. T. and Hall, M. R. (1984): Verborgene Signale, Hamburg.

Hall, E. T. and Hall, M. R. (1990): Understanding Cultural Differences: Germans, French and Americans, Yarmouth.

Hallberg, G. (2004): Is your loyalty programme really building loyalty? Why increasing emotional attachment, not just repeat buying, is key to maximising programme success, in: Journal of Targeting, Measurement & Analysis for Marketing 12(3): 231-241.

Hallowell, R. (1996): The relationships of customer satisfaction, customer loyalty, and profitability: an empirical study, in: International Journal of Service Industry Management 7(4): 27-42.

Harris, L. C. and Goode, M. M. H. (2004): The four levels of loyalty and the pivotal role of trust: a study of online service dynamics, in: Journal of Retailing 80(2): 139-158.

Hellier, P. K., Geursen, G. M., Carr, R. A., and Rickard, J. A. (2003): Customer repurchase intention: A general structural equation model, in: European Journal of Marketing 37(11/12): 1762-1800.

Hennig-Thurau, T. (2004): Customer orientation of service employees: Its impact on customer satisfaction, commitment, and retention, in: International Journal of Service Industry Management 15(5): 460-478.

Henning-Thurau, T., Gwinner, K. P., and Gremler, D. D. (2000): Why Customers Build Relationships with Companies - And Why Not, in: Henning-Thurau, T. and Hansen, U., Relationship marketing: Gaining competitive advantage through customer satisfaction and customer retention, Heidelberg, New York: 369-391.

Heskett, J. L. (1977): Logistics - essential to strategy, in: Harvard Business Review 55(6): 85-96.

Hill, S. (1994): Logistics takes new road, in: Manufacturing Systems November 1994: 28-32.

Hocutt, M. A. (1998): Relationship dissolution model: antecedents of relationship commitment and the likelihood of dissolving a relationship, in: International Journal of Service Industry Management 9: 189-200.

Hofstede, G. (1980): Culture's Consequences: International Differences in Work-Related Values, Beverly Hills.

Hofstede, G. (1983): National Cultures in Four Dimensions: A research-based Theory of Cultural Differences among Nations, in: International Studies of Management & Organization 13(1/2): 46-74.

Hofstede, G. (1985): The Interaction between National and Organizational Value Systems, in: Journal of Management Studies 22(4): 347-357.

Hofstede, G. (1989): Cultural Predictors of National Negotiation Styles, in: Mautner-Markhof, F., Processes on International Negotiations, Boulder: 193-201.

Hofstede, G. (1996): Riding the waves of commerce: a test of Trompenaars' "model" of national culture differences, in: International Journal of Intercultural Relations 20(2): 189-198.

Hofstede, G. (1997): Riding the Waves: A Rejoinder, in: International Journal of Intercultural Relations 21(2): 287-290.

Hofstede, G. (2001): Culture's Consequences: Comparing Values, Behaviors, Institutions, and Organizations Across Nations, 2. Edition, Thousand Oaks.

Homans, G. C. (1958): Social Behavior as Exchange, in: American Journal of Sociology 63(5): 597-606.

Homburg, C. (1989): Exploratorische Ansätze der Kausalanalyse als Instrument der Marketingplanung, Frankfurt am Main.

Homburg, C. and Baumgartner, H. (1995): Beurteilung von Kausalmodellen: Bestandsaufnahme und Anwendungsempfehlungen, in: Marketing - ZFP 17: 162-176.

Homburg, C. and Fassnacht, M. (2001): Kundennähe, Kundenzufriedenheit und Kundenbindung bei Dienstleistungsunternehmen, in: Bruhn, M. and Meffert, H., Handbuch Dienstleistungsmanagement, Wiesbaden: 441-463.

Homburg, C. and Fürst, A. (2005): How Organizational Complaint Handling Drives Customer Loyalty: An Analysis of the Mechanistic and the Organic Approach, in: Journal of Marketing 69(3): 95-114.

Homburg, C. and Giering, A. (1996): Konzeptualisierung und Operationalisierung komplexer Konstrukte: Ein Leitfaden für die Marketingforschung, in: Marketing - ZFP 18(1): 5-24.

Homburg, C. and Giering, A. (1998): Konzeptualisierung und Operationalisierung komplexer Konstrukte: Ein Leitfaden für die Marketingforschung, in: Hildebrandt, L. and Homburg, C., Die Kausalanalyse: Instrument der empirischen betriebswirtschaftlichen Forschung, Stuttgart: 111-148.

Homburg, C., Giering, A., and Menon, A. (2003): Relationship Characteristics as Moderators of the Satisfaction-Loyalty Link: Findings in a Business-to-Business Context, in: Journal of Business-to-Business Marketing 10(3): 35-62.

Hoppe, M. H. (2004): An interview with Geert Hofstede, in: Academy of Management Executive 18(1): 75-79.

Hu, L. and Bentler, P. M. (1995): Evaluating Model Fit, in: Hoyle, R. H., Structural Equation Modeling: Concepts, Issues, and Applications, London, New Delhi: 76-99.

Huang, Y., Rayner, C., and Zhuang, L. (2003): Does intercultural competence matter in intercultural business relationship development?, in: International Journal of Logistics: Research & Applications 6(4): 277-288.

Huff, L. and Kelley, L. (2005): Is collectivism a liability? The impact of culture on organizational trust and customer orientation: a seven-nation study, in: Journal of Business Research 58(1): 96-102.

Huppertz, J. W., Arenson, S. J., and Evans, R. H. (1978): An Application of Equity Theory to Buyer-Seller Exchange Situations, in: Journal of Marketing Research 15(2): 250.

Innis, D. E. and La Londe, B. J. (1994): Customer service: The key to customer satisfaction, customer loyalty, and market share, in: Journal of Business Logistics 15(1): 1-27.

Jacoby, J. and Kyner, D. B. (1973): Brand Loyalty Vs. Repeat Purchasing Behavior, in: Journal of Marketing Research 10(1): 1-9.

Jameson, D. A. (1992): Understanding Cultural Differences/Management in Two Cultures, in: Journal of Business Communication 29(4): 404-407.

Jiang, P. and Rosenbloom, B. (2005): Customer intention to return online: price perception attribution-level performance, and satisfaction unfolding over time, in: European Journal of Marketing 39(1/2): 150-174.

John, G. (1984): An Empirical investigation of Some Antecedents of Opportunism in a Marketing Channel, in: Journal of Marketing Research 21(3): 278-289.

John, G. and Reve, T. (1982): The Reliability and Validity of Key Informant Data from Dyadic Relationships in Marketing, in: Journal of Marketing Research 19: 517-524.

Jöreskog, K. G. and Sörbom, D. (1982): Recent Developments in Structural Equation Modeling, in: Journal of Marketing Research 19(4): 404-416.

Kaiser, H. (1974): An Index of Factorial Simplicity, in: Psychometrika 39 (1984): 31-36.

Kalwani, M. U. and Narayandas, N. (1995): Long-term manufacturer-supplier relationships: Do they pay off for supplier firms?, in: Journal of Marketing 59(1): 1-16.

Kaplan, D. (1995): Statistical Power in Structural Equation Modeling, in: Hoyle, R. H., Structural Equation Modeling: Concepts, Issues, and Applications, London, New Delhi: 100-117.

Keaveney, S. M. (1995): Customer switching behavior in service industries: An exploratory study, in: Journal of Marketing 59(2): 71-82.

Keller, S. B., Savitske, K., Stank, T. P., Lynch, D. F., and Ellinger, A. E. (2002): A Summary of Multi-Item Scales used in Logistics Research, in: Journal of Business Logistics 23: 83-281.

Klaus, P. (1999): Logistik als "Weltsicht", in: Weber, J. and Baumgarten, H., Handbuch Logistik - Management von Material- und Warenflußprozessen, Stuttgart: 15-32.

Kluckhohn, C. (1951): The Study of Culture, in: Lerner, D. and Lasswell, H. D., The Policy Sciences, Stanford: 86-101.

Kluckhohn, F. and Strodtbeck, F. L. (1960): Variations in Value Orientations, Westport.

Knemeyer, A. M., Corsi, T. M., and Murphy, P. R. (2003): Logistics Outsourcing Relationships: Customer Perspectives, in: Journal of Business Logistics 24: 77-109.

Knemeyer, A. M. and Murphy, P. R. (2004): Evaluating the Performance of Third-Party Logistics Arrangements: A Relationship Marketing Perspective, in: Journal of Supply Chain Management 40(1): 35-51.

Knemeyer, A. M. and Murphy, P. R. (2005): Exploring the Potential Impact of Relationship Characteristics and Customer Attributes on the Outcomes of Third-party Logistics Arrangements, in: Transportation Journal 44(1): 5-19.

Kogut, B. and Singh, H. (1988): The Effects of National Culture on the Choice of Entry Mode, in: Journal of International Business Studies 19(3): 411-432.

Kumar, N., Stern, L. W., and Anderson, J. C. (1993): Conducting interorganizational research using key informants, in: Academy of Management Journal 36(6): 1633-1651.

Kumar, V., Bohling, T. R., and Ladda, R. N. (2003): Antecedents and consequences of relationship intention: Implications for transaction and relationship marketing, in: Industrial Marketing Management 32: 667-676.

Kwon, I.-W. G. and Suh, T. (2005): Trust, commitment and relationships in supply chain management: a path analysis, in: Supply Chain Management: An International Journal 10(1): 26-33.

La Londe, B. J. and Maltz, A. B. (1992): Some Propositions About Outsourcing The Logistics Function, in: International Journal of Logistics Management 3(1): 1-11.

Lacity, M. C., Willcocks, L. P., and Feeny, D. F. (1995): IT Outsourcing: Maximize Flexibility and Control, in: Harvard Business Review 73(3): 84-93.

Lam, S. Y., Shankar, V., Erramilli, M. K., and Murthy, B. (2004): Customer Value, Satisfaction, Loyalty, and Switching Costs: An Illustration From a Business-to-Business Service Context, in: Journal of the Academy of Marketing Science 32(3): 293-311.

Lambe, C. J., Wittmann, C. M., and Spekman, R. E. (2001): Social Exchange Theory and Research on Business-to-Business Relational Exchange, in: Journal of Business-to-Business Marketing 8(3): 1-36.

Lambert, D. M., Emmelhainz, M. A., and Gardner, J. T. (1999): Building Successful Logistics Partnerships, in: Journal of Business Logistics 20(1): 165-181.

Lambert, D. M. and Harrington, T. C. (1990): Measuring Nonresponse Bias in Customer Service Mail Surveys, in: Journal of Business Logistics 11(2): 5-25.

Lane, C. (2000): Theories and Issues in the Study of Trust, in: Lane, C. and Bachmann, R., Trust Within and Between Organizations: Conceptual Issues and Empirical Applications, New York: 1-30.

Lane, C. and Bachmann, R. (1996): The Social Constitution of Trust: Supplier Relations in Britain and Germany, in: Organizational Studies 17: 365-395.

Langley, C. J., van Dort, E., Ang, A., and Sykes, S. R. (2005): 2005 Third-Party Logistics: Results and Findings of the 10th Annual Study.

Lapierre, J., Filiatrault, P., and Chebat, J.-C. (1999): Value Strategy Rather Than Quality Strategy: A Case of Business-to-Business Professional Services, in: Journal of Business Research 45(2): 235-246.

Laroche, M., Ueltschy, L. C., Abe, S., Cleveland, M., and Yannopoulos, P. P. (2004): Service Quality Perceptions and Customer Satisfaction: Evaluating the Role of Culture, in: Journal of International Marketing 12(3): 58-85.

Larson, P. D. and Halldorsson, A. (2004): Logistics versus supply chain management: an international survey, in: International Journal of Logistics: Research & Applications 7(1): 17-31.

Larson, P. D. and Poist, R. F. (2004): Improving Response Rates to Mail Surveys: A Research Note, in: Transportation Journal 43(4): 67-74.

Lee, J., Lee, J., and Feick, L. (2001): The impact of switching costs on the customer satisfaction-loyalty link: mobile phone service in France, in: Journal of Services Marketing 15(1): 35-48.

Lemon, K. N., White, T. B., and Winer, R. S. (2002): Dynamic Customer Relationship Management: Incorporating Future Considerations into the Service Retention Decision, in: Journal of Marketing 66: 1-14.

Lewin, K. (1936): Some Social-Psychological Differences between the US and Germany, in: Character and Personality 4: 265-293.

Lewis, I. and Talalayevsky, A. (2000): Third-Party Logistics: Leveraging Information Technology, in: Journal of Business Logistics 21(2): 173-185.

Lewis, M. (2004): The Influence of Loyalty Programs and Short-Term Promotions on Customer Retention, in: Journal of Marketing Research (JMR) 41(3): 281-292.

Lieb, R. C. (1992): The use of third-party logistics services by large American manufacturers, in: Journal of Business Logistics 13(2): 29-42.

Lieb, R. C., Millen, R. A., and van Wassenhove, L. N. (1993): Third-party Logistics Services: A Comparison of Experienced American and European Manufacturers, in: International Journal of Physical Distribution & Logistics 23(6): 35-44.

Lieb, R. C. and Miller, J. (2002): The Use of Third-party Logistics Services by Large US Manufacturers, The 2000 Survey, in: International Journal of Logistics: Research & Applications 5(1): 1-12.

Lieb, R. C. and Randall, H. L. (1999): 1997 CEO Perspectives on the Current Status and Future Prospects of the Third Party Logistics Industry in the United States, in: Transportation Journal 38(3): 28-41.

Liljander, V. and Strandvik, T. (1995): The nature of customer relationships in services, in: Advances in Services Marketing and Management 4: 141-167.

Lin, C.-P. and Ding, C. G. (2005): Opening the black box: Assessing the mediating mechanism of relationship quality and the moderating effects of prior experience in ISP service, in: International Journal of Service Industry Management 16(1): 55-80.

Liu, B. S.-C., Furrer, O., and Sudharshan, D. (2001): The Relationships Between Culture and Behavioral Intentions Toward Services, in: Journal of Service Research 4(2): 118-129.

Luhmann, N. (1973): Vertrauen: Ein Mechanismus der Reduktion sozialer Komplexität, 2. Edition, Stuttgart.

Luk, S. T. K. (1997): An examination of the role of marketing culture in service quality, in: International Journal of Contemporary Hospitality Management 9(1): 13-20.

Lynch, M. E., Imada, S. J., and Bookbinder, J. H. (1994): The future of logistics in Canada: A Delphi-based forecast, in: Logistics & Transportation Review 30(1): 95-112.

Macintosh, G. and Lockshin, L. S. (1997): Retail relationships and store loyalty: A multi-level perspective, in: International Journal of Research in Marketing 14(5): 487-497.

Maltz, A. (1994): Outsourcing the warehousing function: economic and strategic considerations, in: Logistics and Transportation Review 30(3): 245-265.

Marsh, H. W. (1998): Pairwise deletion for missing data in structural equation models: Nonpositive definite matrices, parameter estimates, goodness of fit, and adjusted sample sizes, in: Structural Equation Modeling: A Multidisciplinary Journal 5: 22-36.

Matthes, A. K. (2006): Die Wirkung von Vertrauen auf die Ex-Post-Transaktionskosten in Kooperation und Hierarchie, Forthcoming.

Mattila, A. S. (1999): The role of culture and purchase motivation in service encounter evaluations, in: Journal of Services Marketing 13(4/5): 376-389.

McCabe, J. V. (1990): Outside Managers Offer Packaged Export Expertise, in: Journal of Business Strategy 11(2): 20-23.

McCort, D. J. and Malhotra, N. K. (1993): Culture and consumer behavior: Toward an understanding of cross-cultural consumer behavior in international marketing, in: Journal of International Consumer Marketing 6(2): 91-127.

McGee, G. W. and Ford, R. C. (1987): Two (or More?) Dimensions of Organizational Commitment: Reexamination of the Affective and Continuance Commitment Scales, in: Journal of Applied Psychology 72(4): 638-642.

McGinnis, M. A., Kochunny, C. M., and Ackerman, K. B. (1995): Third Party Logistics Choice, in: The International Journal of Logistics Management 6: 93-102.

McIntosh, S. S. (1999): Defining Cultures, in: HR Magazine 44(4): 144-147.

McIvor, R. (2000): A practical framework for understanding the outsourcing process, in: Supply Chain Management 5(1): 22-36.

McSweeney, B. (2002): Hofstede's Model of National Cultural Differences and Their Consequences: A Triumph of Faith - A Failure of Analysis, in: Human Relations 55(1): 89-118.

Mendonsa, R. M. (1999): Riding the Waves of Culture: Understanding Cultural Diversity in Business, in: Human Resource Development Quarterly 10(2): 185-188.

Mentzer, J. T. and Flint, D. J. (1997): Validity in Logistics Research, in: Journal of Business Logistics 18(1): 199-216.

Mentzer, J. T., Flint, D. J., and Kent, J. L. (1999): Developing a Logistics Service Quality Scale, in: Journal of Business Logistics 20(1): 9-32.

Mentzer, J. T. and Kahn, K. B. (1995): A framework of logistics research, in: Journal of Business Logistics 16(1): 231-250.

Meyer, A. and Oevermann, D. (1995): Kundenbindung, in: Tietz, B., Köhler, R., and Zentes, J., Handbuch des Marketing, 2. Edition, Stuttgart: 1340-1351.

Meyer, J. P. and Allen, N. J. (1984): Testing the 'Side-Bet Theory' of Organizational Commitment: Some Methodological Considerations, in: Journal of Applied Psychology 69(3): 372-378.

Mittal, V. and Kamakura, W. A. (2001): Satisfaction, Repurchase Intent, and Repurchase Behavior: Investigating the Moderating Effect of Customer Characteristics, in: Journal of Marketing Research 38(1): 131-142.

Mittal, V., Kumar, P., and Tsiros, M. (1999): Attribute-Level Performance, Satisfaction, and Behavioral Intentions over Time: a Consumption-System Approach, in: Journal of Marketing 63(2): 88-101.

Mittal, V., Ross, W. T., and Baldasare, P. M. (1998): The Asymmetric Impact of Negative and Positive Attribute-Level Performance on Overall Satisfaction and Repurchase Intentions, in: Journal of Marketing 62: 33-47.

Molm, L. D., Takahashi, N., and Petersen, G. (2000): Risk and Trust in Social Exchange: An Experimental Test of a Classical Proposition, in: American Journal of Sociology 105: 1396-1427.

Money, R. B. (2004): Word-of-mouth promotion and switching behavior in Japanese and American business-to-business service clients, in: Journal of Business Research 57(3): 297-305.

Moore, K. R. (1998): Trust and relationship commitment in logistics alliances: A buyer perspective, in: International Journal of Purchasing & Materials Management 34(1): 24-37.

Moore, K. R. and Cunningham III, W. A. (1999): Social exchange behavior in logistics relationships: a shipper perspective, in: International Journal of Physical Distribution & Logistics 29: 103-121.

Moorman, C., Deshpandé, R., and Zaltman, G. (1993): Factors Affecting Trust in Market Research Relationships, in: Journal of Marketing 57: 81-101.

Moorman, C., Zaltman, G., and Deshpandé, R. (1992): Relationships between Providers and Users of Marketing Research: The Dynamics of Trust within and between Organizations, in: Journal of Marketing Research 29(8): 314-329.

Morain, G. (1991): Book Review: Understanding Cultural Differences: Germans, French and Americans, in: The Modern Language Journal(75): 135-136.

Morash, E. A. and Clinton, S. R. (1997): The Role of Transportation Capabilities in International Supply Chain Management, in: Transportation Journal 36(3): 5-17.

Morgan, R. M. and Hunt, S. D. (1994): The Commitment-Trust Theory of Relationship Marketing, in: Journal of Marketing 58(3): 20-38.

Morris, D. S., Barnes, B. R., and Lynch, J. E. (1999): Relationship marketing needs total quality management, in: Total Quality Management 10(4/5): S659.

Mulaik, S. A. (1987): A Brief History of the Philosophical Foundations of Exploratory Factor Analysis, in: Multivariate Behavioral Research 22(3): 267-305.

Mulaik, S. A. and James, L. R. (1995): Objectivity and Reasoning in Science and Structural Equation Modeling, in: Hoyle, R. H., Structural Equation Modeling: Concepts, Issues, and Applications, London, New Delhi: 118-137.

Muller, E. J. (1993): The top guns of third party logistics, in: Distribution 92(3): 30-38.

Murphy, P. R., Daley, J. M., and Knemeyer, A. M. (1999): Comparing Logistics Management in Small and Large Firms: An Exploratory Study, in: Transportation Journal 38(4): 18-25.

Murphy, P. R. and Poist, R. F. (2003): Green perspectives and practices: a "comparative logistics" study, in: Supply Chain Management 8: 122-131.

Narayandas, D. and Rangan, K. (2004): Building and Sustaining Buyer-Seller Relationships in Mature Industrial Markets, in: Journal of Marketing 68: 63-77.

Neal, W. D. (1999): Satisfaction is Nice, But Value Drives Loyalty, in: Marketing Research 11(1): 20-23.

Neuert, J., Butin, J., Farfelan, A.-S., Kolar, P., and Redlich, T. (2005): Rational Conduct, Fairness, and Reciprocity in Economic Transaction Processes, in: Journal of American Academy of Business, Cambridge 7(1): 338-353.

Ngobo, P. V. (2004): Drivers of customers' cross-buying intentions, in: European Journal of Marketing 38(9/10): 1129-1157.

Nunnally, J. C. (1967): Psychometric Theory, New York.

Oliver, R. L. (1997): Satisfaction: A behavioral perspective on the consumer, New York.

Oliver, R. L., Rust, R. T., and Varki, P. (1997): Customer Delight: Foundations, Findings, and Managerial Insight, in: Journal of Retailing 73(3): 311-336.

Olsen, S. O. (2002): Comparative Evaluation and the Relationship Between Quality, Satisfaction, and Repurchase Loyalty, in: Journal of the Academy of Marketing Science 30(3): 240-249.

Ornauer, H., Wiberg, H., Sicinski, A., and Galtung, J. (1976): Images of the world in the year 2000: A comparative ten-nation study, The Hague.

Panayides, P. (2004): Logistics Service Providers: An Empirical Study of Marketing Strategies and Company Performance, in: International Journal of Logistics 7(1): 1-15.

Parasuraman, A., Zeithaml, V. A., and Berry, L. L. (1985): A Conceptual Model of Service Quality and Its Implications for Future Research, in: Journal of Marketing 49(4): 41-50.

Parasuraman, A., Zeithaml, V. A., and Berry, L. L. (1988): SERVQUAL: A Multiple-Item Scale for Measuring Consumer Perceptions of Service Quality, in: Journal of Retailing 64(1): 12-40.

Parsons, T. (1951): The Social System, New York.

Persson, G. and Virum, H. (2001): Growth Strategies for Logistics Service Providers: A Case Study, in: International Journal of Logistics Management 12(1): 53-64.

Peter, J. P. (1979): Reliability: A Review of Psychometric Basics and Recent Marketing Practices, in: Journal of Marketing Research (JMR) 16(1): 6-17.

Peter, J. P. (1981): Contruct Validity: A Review of Basic Issues and Marketing Practices, in: Journal of Marketing Research 18: 133-145.

Peter, J. P. and Churchill Jr., G. A. (1986): Relationships Among Research Design Choices and Psychometric Properties of Rating Scales:

A Meta-Analysis, in: Journal of Marketing Research (JMR) 23(1): 1-10.

Peter, S. (1997): Kundenbindung als Marketingziel, Wiesbaden.

Pfohl, H.-C. and Zöllner, W. (1997): Organization for logistics: The contingency approach, in: International Journal of Physical Distribution & Logistics Management 27(5/6): 306-320.

Ping, R. A. J. (1993): The Effects of Satisfaction and Structural Constraints on Retailer Exiting, Voice, Loyalty, Opportunism, and Neglect, in: Journal of Retailing 69(3): 320-352.

Pressey, A. D. and Selassie, H. G. (2003): Are cultural difference overrated? Examining the influence of national culture on international buyer-seller relationships, in: Journal of Consumer Behaviour 2(4): 354-368.

Pritchard, M. P., Havitz, M. E., and Howard, D. R. (1999): Analyzing the Commitment-Loyalty Link in Service Contexts, in: Journal of the Academy of Marketing Science 27(3): 333-348.

Puck, J. F. (2002): Convergence and Divergence of Management Practices: The Case of Corporate Web Site Recruiting, Working Paper 4/2002, Chair of International Management, Nuremberg.

Rabinovich, E., Windle, R., Dresner, M., and Corsi, T. M. (1999): Outsourcing of integrated logistics functions - An examination of industry practices, in: International Journal of Physical Distribution & Logistics Management 29(6): 353-373.

Ranaweera, C. and Neely, A. (2003): Some moderating effects on the service quality-customer retention link, in: International Journal of Operations & Production Management 23(2): 230-248.

Ranaweera, C. and Prabhu, J. (2003): The influence of satisfaction, trust and switching barriers on customer retention on a continuous purchasing setting, in: International Journal of Service Industry Management 14(4): 374-395.

Rao, K., Young, R. R., and Novick, J. A. (1993): Third party services in the logistics of global firms, in: Logistics & Transportation Review 29(4): 363-370.

Razzaque, M. A. and Cheng, C. C. (1998): Outsourcing of logistics functions: A literature survey, in: International Journal of Physical Distribution & Logistics Management 28(2-4): 89-107.

Reichheld, F. F., Markey, R. G. J., and Hopton, C. (2000): The loyalty effect - the relationship between loyalty and profits, in: European Business Journal 12(3): 134-139.

Reichheld, F. F. and Sasser, W. C. J. (1990): Zero defections: Quality comes to services, in: Harvard Business Review 68(5): 105-111.

Reinartz, W., Thomas, J. S., and Kumar, V. (2005): Balancing Acquisition and Retention Resources to Maximize Customer Profitability, in: Journal of Marketing 69: 63-79.

Reinecke, J. (2003): Strukturgleichungsmodelle mit latenten Variablen: Meßmodelle und konfirmatorische Faktorenanalyse, Working Paper, Universität Trier.

Reynolds, K. E. and Beatty, S. E. (1999): Customer Benefits and Company Consequences of Customer-Salesperson Relationships in Retailing, in: Journal of Retailing 75(1): 11-32.

Richardson, H. L. (1990): Explore outsourcing, in: Transportation & Distribution 31(7): 17-20.

Richardson, H. L. (1992): Outsourcing: the power worksource, in: Transportation & Distribution 33(7): 22-24.

Richardson, H. L. (1993): Why use third parties?, in: Transportation & Distribution 34(1): 29-31.

Richardson, H. L. (1995): Logistics help for the challenged, in: Transportation & Distribution 36(1): 60-74.

Rindfleisch, A. and Heide, J. B. (1997): Transaction cost analysis: Past, present, and future applications, in: Journal of Marketing 61(4): 30-54.

Rodriguez, C. M. and Wilson, D. T. (2002): Relationship Bonding and Trust as a Foundation for Commitment in U.S.-Mexican Strategic Alliances: A Structural Equation Modeling Approach, in: Journal of International Marketing 10(4): 53-76.

Roehm, M. L., Pullins, E. B., and Roehm Jr., H. A. (2002): Designing Loyalty-Building Programs for Packaged Goods Brands, in: Journal of Marketing Research 39(2): 202-213.

Rust, R. T. and Zahorik, A. J. (1993): Customer Satisfaction, Customer Retention, and Market Share, in: Journal of Retailing 69(2): 193-215.

Sako, M. (2000): Does Trust Improve Business Performance?, in: Lane, C. B., R., Trust Within and Between Organizations: Conceptual Issues and Empirical Applications, New York: 88-117.

Samiee, S. and Walters, P. G. P. (2003): Relationship marketing in an international context: a literature review, in: International Business Review 12: 193-214.

Sattler, H., Schrader, S., and Lüthje, C. (2003): Informal cooperation in the US and Germany: cooperative managerial capitalism vs. competitive managerial capitalism in interfirm information trading, in: International Business Review 12(3): 273-295.

Schäfer-Kunz, J. and Tewald, C. (1998): Make-or-buy Entscheidungen in der Logistik, Wiesbaden.

Selnes, F. (1993): An Examination of the Effect of Product Performance on Brand Reputation, Satisfaction and Loyalty, in: European Journal of Marketing 27(9): 19-35.

Shane, S. (1994): Uncertainty Avoidance and the Preference for Innovation Championing Roles, in: Journal of International Business Studies First Quarter: 47-68.

Shankar, V., Smith, A. K., and Rangaswamy, A. (2003): Customer satisfaction and loyalty in online and offline environments, in: International Journal of Research in Marketing 20(2): 153-175.

Sharma, N. and Patterson, P. G. (1999): The impact of communication effectiveness and service quality on relationship commitment in consumer, professional services, in: The Journal of Services Marketing 13: 151-170.

Sharp, B. and Sharp, A. (1997): Loyalty programs and their impact on repeat-purchase loyalty patterns, in: International Journal of Research in Marketing 14(5): 473-486.

Sheffi, Y. (1990): Third party logistics: Present and future prospects, in: Journal of Business Logistics 11(2): 27-39.

Shemwell, D. J., Cronin, J. J., and Bullard, W. R. (1994): Relational Exchange in Services: An Empirical Investigation of Ongoing Customer Service-provider Relationships, in: International Journal of Service Industry Management 5: 57-68.

Siems, F. (2003): Preiswahrnehmung von Dienstleistungen: Konzeptualisierung und Integration in das Relationship Marketing, Wiesbaden.

Simon, H. (1988): Wettbewerbsvorteile und Wettbewerbsfähigkeit, Stuttgart.

Singh, N. (2004): From Cultural Models to Cultural Categories: A Framework for Cultural Analysis, in: Journal of American Academy of Business, Cambridge 5(1/2): 95-101.

Sirdeshmukh, D., Singh, J., and Sabol, B. (2002): Consumer Trust, Value, and Loyalty in Relational Exchanges, in: Journal of Marketing 66: 15-37.

Sirohi, N., McLaughlin, E. W., and Wittink, D. R. (1998): A Model of Consumer Perceptions and Store Loyalty Intentions for a Supermarket Retailer, in: Journal of Retailing 74(2): 223-245.

Sivadas, E. and Baker-Prewitt, J. L. (2000): An examination of the relationship between service quality, customer satisfaction, and store loyalty, in: International Journal of Retail & Distribution Management 28(2/3): 73-82.

Skjoett-Larsen, T. (2000): Third party logistics-from an interorganizational point of view, in: International Journal of Physical Distribution & Logistics Management 30(2): 112-127.

Slangen, A. and Hennart, J.-F. (2003): Cultural Distance and Foreign Direct Investment: A Comprehensive Model Explaining the Impact of National Cultural Differences on Entry Mode and Choice and Subsidiary Performance. 28th Annual Meeting of the European International Business Academy, Athens.

Sohal, A. S., Millen, R., Maggard, M., and Moss, S. (1999): Quality in Logistics: a comparison of practices between Australian and North American/European firms, in: International Journal of Physical Distribution & Logistics 29: 267-280.

Spreng, R. A., MacKenzie, S. B., and Olshavsky, R. W. (1996): A Reexamination of the Determinants of Consumer Satisfaction, in: Journal of Marketing 60: 15-32.

Sriram, V. and Mummalaneni, V. (1990): Determinants of Source Loyalty in Buyer-Seller Relationships, in: Journal of Purchasing & Materials Management 26(1): 21-26.

Stank, T. P., Goldsby, T. J., and Vickery, S. K. (1999): Effect of service supplier performance on satisfaction and loyality of store managers in the fast food industry, in: Journal of Operations Management 17: 429-447.

Stank, T. P., Goldsby, T. J., Vickery, S. K., and Savitskie, K. (2003): Logistics Service Performance: Estimating its Influence on Market Share, in: Journal of Business Logistics 24(1): 27-55.

Steenkamp, J.-B. and Baumgartner, H. (1998a): Assessing measurement invariance in cross-national consumer research, in: Journal of Consumer Research 25(1): 78-90.

Steenkamp, J.-B. and Baumgartner, H. (1998b): Assessing Measurement Invariance through Multi-Sample Structural Equation Modeling, in: Hildebrandt, L., Homburg, C., Die Kausalanalyse: Ein Instrument der empirischen betriebswirtschaftlichen Forschung, Stuttgart: 399-426.

Stone, M., Woodcock, N., and Wilson, M. (1996): Managing the Change from Marketing Planning to Customer Relationship Management, in: Long Range Planning 29(5): 675-683.

Stone, M. A. (2001): European Expansion of UK Third-party Logistics Service Providers, in: International Journal of Logistics: Research & Applications 4(1): 97-115.

Sultan, F. and Simpson Jr., M. C. (2000): International service variants: airline passenger expectations and perceptions of service quality, in: Journal of Services Marketing 14(2/3): 188-216.

Sum, C.-C. and Teo, C.-B. (1999): Strategic posture of logistics service providers in Singapore, in: International Journal of Physical Distribution & Logistics Management 29(9/10): 588-605.

Sureshchandar, G. S., Rajendran, C., and Anantharaman, R. N. (2002): The relationship between service quality and customer satisfaction - a factor specific approach, in: Journal of Services Marketing 16: 363-379.

Sydow, J. (2000): Understanding the Constitution of Interorganizational Trust, in: Lane, C. B., R., Trust Within and Between Organizations: Conceptual Issues and Empirical Applications, New York: 31-63.

Taylor, S. A. and Baker, T. L. (1994): An assessment of the relationship between service quality and customer satisfaction in the Formation on Consumers' Purchase Intentions, in: Journal of Retailing 70(2): 163-178.

Thelen, S. T. and Honeycutt Jr., E. D. (2004): Assessing National Identity in Russia Between Generations Using the National Identity Scale, in: Journal of International Marketing 12(2): 58-81.

Thibaut, J. W. and Kelley, H. H. (1959): The Social Psychology of Groups, New York, London, Sydney.

Thompson, L. and Spanier, G. (1983): The End of Marriage and Acceptance of Marital Termination, in: Journal of Marriage and the Family 45: 103-113.

Too, L. H. Y., Souchon, A. L., and Thirkell, P. C. (2001): Relationship Marketing and Customer Loyalty in a Retail Setting: A Dyadic Exploration, in: Journal of Marketing Management 17: 287-319.

Trompenaars, F. (1996): Resolving International Conflict: Culture and Business Strategy, in: Business Strategy Review 7(3): 51-68.

Trompenaars, F. and Hampden-Turner, C. (1997): Riding the Waves of Culture: Understanding Cultural Diversity in Business, 2. Edition, London, Naperville.

Trunick, P. A. (1989): Outsourcing: a single source for many talents, in: Transportation & Distribution 30(7): 20-23.

Trunick, P. A. (1992): Carving a niche in global logistics, in: Transportation & Distribution 33(2): 57-58.

Ueltschy, L. C. and Krampf, R. F. (2001): Cultural Sensitivity to Satisfaction and Service Quality Measures, in: Journal of Marketing 9(3): 14-31.

van Dyke, T. P., Prybutok, V. R., and Kappelmann, L. A. (1999): Cautions on the Use of the SERVQUAL Measure to Assess the Quality of Information Systems Services, in: Decision Sciences 30: 877-891.

van Hoek, R. I. (2000): The Purchasing and Control of Supplementary Third-Party Logistics Services, in: Journal of Supply Chain Management: A Global Review of Purchasing & Supply 36(4): 14-26.

van Laarhoven, P., Berglund, M., and Peters, M. (2000): Third-party logistics in Europe - five years later, in: International Journal of Physical Distribution & Logistics Management 30(5): 425-442.

Varki, S. and Colgate, M. (2001): The Role of Price Perceptions in an Integrated Model of Behavioral Intentions, in: Journal of Service Research 3(3): 232-240.

Venetis, K. A. and Ghauri, P. N. (2004): Service quality and customer retention: building long-term relationships, in: European Journal of Marketing 38(11/12): 1577-1598.

Verhoef, P. C. (2003): Understanding the Effect of Customer Relationship Management Efforts on Customer Retention and Customer Share Development, in: Journal of Marketing 67(4): 30-45.

Voss, G. B., Parasuraman, A., and Grewal, D. (1998): The Roles of Price, Performance, and Expectations in Determining Satisfaction in Service Exchanges, in: Journal of Marketing 62(4): 46-61.

Wallace, D. W., Giese, J. L., and Johnson, J. L. (2004): Customer retailer loyalty in the context of multiple channel strategies, in: Journal of Retailing 80(4): 249-263.

Wallenburg, C. M. (2003): Kundenbindung in der Logistik: Eine empirische Untersuchung zu Geschäftsbeziehungen zwischen Logistikdienstleistern und ihren Kunden, in: Weber, J. and Deepen, J., Erfolg durch Logistik: Erkenntnisse aktueller Forschung, Bern, Stuttgart, Vienna: 77-114.

Wallenburg, C. M. (2004): Kundenbindung in der Logistik: Eine empirische Untersuchung zu ihren Einflussfaktoren, Bern, Stuttgart, Vienna.

Wallenburg, C. M. and Weber, J. (2005): Structural Equation Modelling as a Basis for Theory Development within Logistics and Supply Chain Management Research, in: Kotzab, H., Seuring, S., Müller, M., and Reiner, G., Research Methodologies in Supply Chain Management, Heidelberg, New York: 171-186.

Wathne, K. H. and Heide, J. B. (2000): Opportunism in Interfirm Relationships: Forms, Outcomes, and Solutions, in: Journal of Marketing 64(4): 36-51.

Weber, J. (2002): Logistikkostenrechnung, 2. Edition, Berlin.

Weber, J. and Dehler, M. (1999): Erfolgsfaktor Logistik - Wunsch und Wirklichkeit, in: Logistik Heute 24(9): 34-41.

Weber, J. and Kummer, S. (1998): Logistikmanagement, Stuttgart.

Weber, J. and Wallenburg, C. M. (2004): Zusatzbeauftragung von Logistikdienstleistern: empirische Ergebnisse und konzeptionelle Überlegungen zu entsprechenden Defiziten, in: Logistik Management 6(3): 34-46.

Wentworth, F. (2003): Outsourcing Services: The Case Against, in: Logistics & Transport Focus 5(2): 57-59.

Werner, H. (1997): Relationales Beschaffungsverhalten: Ausprägungen und Determinanten, Wiesbaden.

Westbrook, R. A. and Oliver, R. L. (1991): The Dimensionality of Consumption Emotion Patterns and Consumer Satisfaction, in: Journal of Consumer Research 18(1): 84-91.

Wetzels, M. G. M., de Ruyter, J. C., and van Birgelen, M. J. H. (1998): Marketing service relationships: the role of commitment, in: Journal of Business & Industrial Marketing 13: 406-423.

Whybark, D. C. (1990): Education and Global Logistics, in: Logistics and Transportation Review 26(3): 261-270.

Willauer, B. (2005): Consensus as Key Success Factor in Strategy-Making, Wiesbaden.

Williams, J. D., Han, S.-L., and Qualls, W. J. (1998): A Conceptual Model and Study of Cross-Cultural Business Relationships, in: Journal of Business Research 42(2): 135-143.

Williamson, O. E. (1975): Markets and Hierachies: Analysis and Antitrust Implications, New York.

Williamson, O. E. (1985): The Economic Institutions of Capitalism: Firms, Markets, Relational Contracting, New York.

Williamson, O. E. (1990): Transaction Cost Economics, in: R., S. and Willig, R. D., Handbook of Industrial Organization, Amsterdam, New York, Oxford, Tokyo.

Wilson, J. R. (1994): Word-of-Mouth Marketing, New York.

Wilson, W. R. and Peterson, R. E. (1989): Some Limits on the Potency of Word-Of-Mouth Information, in: Advances in Consumer Research 16(1): 23-29.

Winsted, K. F. (1999): Evaluating Service Encounters: A Cross-Cultural and Cross-Industry Exploration, in: Journal of Marketing Theory & Practice 7(2): 106-123.

Witkowski, T. H. and Wolfinbarger, M. F. (2002): Comparative service quality: German and American ratings across service settings, in: Journal of Business Research 55(11): 875-881.

Wong, A. and Sohal, A. (2002): An examination of the relationship between trust, commitment and relationship quality, in: International Journal of Retail & Distribution Management 30(1): 34-50.

Xia, L., Monroe, K. B., and Cox, J. L. (2004): The Price Is Unfair! A Conceptual Framework of Price Fairness Perceptions, in: Journal of Marketing 68(4): 1-15.

Yamagishi, T., Cook, K. S., and Watabe, M. (1998): Uncertainty, Trust, and Commitment Formation in the United States and Japan, in: American Journal of Sociology 104(1): 165-194.

Yammarino, F. J., Skinner, S. J., and Childers, T. L. (1991): Understanding Mail Survey Response Behavior, in: Public Opinion Quarterly 55(4): 613-639.

Yen, H. J. R. and Gwinner, K. P. (2003): Internet retail customer loyalty: the mediating role of relational benefits, in: International Journal of Service Industry Management 14(5): 483-500.

Yi, Y. and Jeon, H. (2003): Effects of Loyalty Programs on Value Perception, Program Loyalty, and Brand Loyalty, in: Journal of the Academy of Marketing Science 31(3): 229-240.

Yim, C. K. and Kannan, P. K. (1999): Consumer Behavioral Loyalty: A Segmentation Model and Analysis, in: Journal of Business Research 44(2): 75-92.

Zeithaml, V. A., Berry, L. L., and Parasuraman, A. (1996): The behavioral consequences of service quality, in: Journal of Marketing 60(2): 31-46.

Zhu, J., Lean, H. S., and Ying, S. K. (2002): The Third-Party Logistics Services and Globalization of Manufacturing, in: International Planning Studies 7(1): 89-104.

Zimmer, P. (2000): Commitment in Geschäftsbeziehungen, Wiesbaden.

Printing: Krips bv, Meppel
Binding: Stürtz, Würzburg